Well Connected

Water and Society
Wendy Jepson, Amber Wutich, and Alexandra Brewis, Series Editors

Well Connected

Everyday Water Practices in Cairo

Tessa Farmer

Johns Hopkins University Press
Baltimore

Printed in the United States of America on acid-free paper
9 8 7 6 5 4 3 2 1

Johns Hopkins University Press
2715 North Charles Street
Baltimore, Maryland 21218
www.press.jhu.edu

Cataloging-in-Publication Data is available from the Library of Congress.
A catalog record for this book is available from the British Library.
ISBN: 978-1-4214-4548-9 (paperback)
ISBN: 978-1-4214-4549-6 (ebook)

Special discounts are available for bulk purchases of this book. For more information, please contact Special Sales at specialsales@jh.edu.

Contents

Acknowledgments

I have accumulated a long list of debts over the course of working on this book, and I owe many thanks. A deep and inadequate thank you goes to the people in Ezbet Khairallah who opened their homes and lives to me. Your names are disguised in this book to protect your anonymity, but each of you knows the truth of your contributions. Your world is rich and beautiful, and I'm grateful that you welcomed me into it.

In Cairo there were countless people who provided intellectual, social, and material support. Dear friends Ibrahim and Omyama, and Mustafa and Regina opened their homes and shared their perspectives on life in Cairo. The lovely Shams Labib Sami always reminded me of the value of commitment, and I have been grateful to catch any time I can with her. Martina Rieker went out of her way to offer support and generously made her connections at the American University in Cairo (AUC) available to me. Mark Westmoreland was a gracious "elder brother" both in Cairo and in Austin before that. A special thanks to Mark's SOAN MA Roundtable at AUC for giving me the opportunity to think through my research material while in the field.

Rania El Essawi came in the guise of a research contact and has remained as a friend and hopeful voice for a better future for Egypt. I extend sincere thanks to the staffs of Anbar Moussa; the Cairo Water Company; the Ain el-Sira and Abaseya water offices; the Ministry of Housing, Utilities and Urban Development; the Holding Company for Water and Wastewater; GTZ; and the Arab contractors for taking the time to speak with me and make documents available. Ghassan Naked, David Osgood, and Kathleen Sheridan from Chemonics were particularly helpful, as were Atef Abdel Sayed at USAID, Mervat Nessiem at the Center for Development Services, and Ali El Faramwy of ISDF. Thank you also to Dr. Magdy Allam for explaining electoral politics in Cairo to me. I'm deeply grateful to be building an ongoing collaboration with May al-Ibrashy and Brook Muller over our shared interest in Cairo's *sabīls*.

Thank you to my friend and dissertation advisor Kamran Asdar Ali, who can never sit still for a compliment but whose passion for critical thought is rivaled only by his compassion for the humanity of his students. His is the voice in my head, urging me to question more and engage more deeply. I don't know what he saw in me as an awkward undergraduate, but I will be forever grateful for the way he has shaped my life. In the deepest sense, I would not be where I am or who I am without him. To my daughter, Maryam, he is Nana Kamran; to me, he is a second father. Syema Muzzaffar and Ehsan and Raavi Ali, thank you for making Austin a wonderful place to be. I can only pay forward your many kindnesses.

Thank you to Katie Stewart, who gives us all permission to listen to our voice as authors and to find what interests us as thinkers; to Polly Strong, who provides thorough and sage advice and endless support; and to Kaushik Ghosh, who reminds me that the best way to read new material is to see how it can expand our vision of the world. My time at the University of Texas at Austin was made immeasurably better by the intellectual tutelage of David Eaton, Charlie Hale, John Hartigan, Elizabeth Keating, Sofian Merabet, Jemima Pierre, and Sam Wilson, along with organizational support from Adriana Dingman and Heather Nathanson-Flowers.

Other friends and colleagues at UT were great compatriots on the journey of scholarly training. In particular, Alisa Perkins was a constant source of thoughtful feedback and exceptional support, as were Somy Kim and Anthony Melvin. I owe thanks to my partner in crime in Cairo and beyond, Johanna Sellman, for renewing my hope in the academic endeavor more often than was fair to ask. Thank you also to Ryan Skinner and Elias and Nils Sellman-Skinner for sharing adventures at home and abroad. To both Naminata Diabate, who thinks critically but acts kindly, and Michal Raizen, who is sensitive to the many registers in which theory operates, thank you for being personal and intellectual role models. Gratitude also goes to Kiran Ahmed, who always shows up with twice what she's received and who ended up feeling more like a mentor to me than a mentee. I appreciate in perpetuity the friendship of Rachid Ait Belhaj, Salih Can Açiksöz, Jenny Carlson, Abdul Haque Chang, Maria Elena Cruz, Alex Dodson, Emily Dylla, Mahyar Entezari, Simeon and Rosie Floyd, Eunice Garza, Juli Grigsby, Hanan Hammad, Sarah Ihmoud, Maryam Kashani, Nedra Lee, Elizabeth LeFlore, Saikat Maitra, Nikki Marterre, Alyssa Miller, Courtney Morris, Mubbashir Rizvi, Jordan and Issam Srour, Heather Teague, Carrie Veilleux, Derrick Washington, Scott and Jen Webel, and Chelsi West Ohueri. Thank you also to the folks I met through the Humanities Institute: Charlotte Canning, Evan Carton, Gretchen Heaton, and Paul Michels.

Major intellectual contributions to this work have come from my collaborators in the Middle East Environmental Worlds writing group. Jessica Barnes, Simone Popperl, Kali Rubaii, Caterina Scaramelli, and Sophia Stamatopoulou-Robbins read each chapter of this book during its long incubation, and I am lucky that some of their glow has reflected onto this project. The intellectual and personal commitment of these scholars has made the academy a much kinder, more collaborative, and more constructive place to be. I can see the imprint of each of them in this manuscript. Kali Rubaii taught me how generative it can be to hold ideas in concert between two minds, and the friendship and accountability that she provided as a book buddy was invaluable. Desh Girod generously read the nearly complete manuscript and offered key insights into major structural questions in the text. Amanda Mabry-Flynn, Magdalena Martinez, and Victoria Reyes have been fabulous coconspirators in our FSP group for years! The work that I did with Katelyn Knox and others through the dissertation-to-book working group taught me how to navigate between the different scales of an argument, from chapter to the overarching takeaway.

Many people have made the University of Virginia a rich intellectual and professional space. Richard Handler has provided astute guidance, honest feedback, and ceaseless advocacy on my behalf. Ira Bashkow, Phoebe Crisman, David Edmunds, Nizar Hermes, Dan Lefkowitz, Steve Parks, and Geeta Patel have all taken me under their wings as mentors and friends. Thank you for making UVA feel like home. Conversations with China Scherz and Sylvia Tidey have been major turning points in the framing of this work. My colleagues in the Global Studies Program and the MESLAC Department deserve a big thank you: Ruth Bernheim, Griff Chaussee, Chris Colvin, John Echeverri-Gent, Caterina Eubanks, Pete Furia, Zvi Gilboa, Laura Goldblatt, Bob Hueckstedt, Gabi Kruks-Wisner, Cliff Maxwell, Farzaneh Milani, Mahshad Mohit, Shankar Nair, Abdul Nasir, Brian Owensby, Kathryn Quissell, and Samhita Sunya, among others. The Department of Anthropology has been a generative intellectual home at UVA; I offer thanks particularly to Jim Igoe, Kasey Jernigan,Patty Wattenmaker, and Kath Weston. Colleagues Fahad Bishara, Feyza Burak Adli, John Casteen, Jim Coan, Peter Debaere, Hope Fitzgerald, Chris Gratien, Ghazal Jafari, Mary Kuhn, David Singerman, and Caitlin Wiley have enriched my experience in countless ways. Ingrid Hakala and Helena Zeweri have been intellectual and personal connections that carry across continents. Thank you to AD's Alison Levine, who shares my interest in pedagogy, and Christian McMillen, who shares my interest in water.

This research has received generous support from a number of institutions, and I am grateful for their faith in my project. The Fulbright-Hays Program offered field research funds and, more than that, provided a uniquely robust in-country resource for all Fulbright scholars. I am indebted to my Fulbright advisor, Hani Moubasher, and his wife, Liz, for their kindness and hospitality while in Cairo; life there would have been much harder without them. The staff of the Fulbright office in Cairo could not have worked harder on our behalf; Hend Ghobrial, Noha El Gindi, and Bruce and Annemarie Lohof all are the tops. The Social Science Research Council (SSRC) has taken a chance on me twice now, first with a dissertation proposal development fellowship (DPDF) in 2007 and later with an international dissertation research fellowship for fieldwork in 2009.

The field directors for the DPDF field of water sustainability, Steve Caton and Ben Orlove, have been generous intellectual guides and have created around them an interdisciplinary cadre of scholars of which I am truly privileged to be part. Caton, who knows water and the Middle East well, agreed to take on my project and has provided smart insights and great kindness. The expanding circle of connections stemming from the SSRC and the Harvard water conference series has been among the most rewarding intellectual spaces that I've encountered, and I'm happy to now count Amahl Bishara, Lisa Björkman, Ashley Carse, Matthew Ellis, Natalia Gutkowski, Mandana Limbert, Emily McKee, Hao Nguyen, Lisa Pfeiffer, Julio Postigo, Sandra Ruckstuhl, Ajantha Subramanian, Sarah Wise, and Rebecca Woods as friends and interlocutors. Special acknowledgments go out to Maya Peterson, who was a brilliant scholar who left us far too soon; and to Martha Kaplan, who shares my passion for the social lives of water coolers, but whose insight is something I can only hope to achieve after a lot of diligent work. Yasmine Moll has become a dear friend and colleague, and her insights into Egypt are endlessly useful.

Thanks also to Carole Alsup and her colleagues at the Philanthropic Educational Organization for their support of this project. Additional support was generously provided by the University of Texas at Austin, the Dean of Arts and Sciences Office at the University of Virginia, the Center for Global Inquiry and Innovation, and the Mellon humanities fellowship through the Institute of the Humanities and Global Culture at UVA.

Scholars in other institutions in the United States have also been sources of rich intellectual and moral support during conferences and workshops, while working on the MES board, and in other collaborations. Lara Deeb and Jessica Winegar have provided excellent examples of what an engaged scholar looks like, and I'm grateful for their guidance and generosity. Thanks to Fida Adely, Lori Allen, Andrea Ballestero, Souad Eddouada, Sarah El-Kazaz, Ilana Feldman, Farha Ghannam, Gökçe Günel, Sondra Hale, Sherine Hamdy, Richard Kernaghan, Petra Kuppinger, Darryl Li, Anne Meneley, Kristin Monroe, Nada Moumtaz, Joanne Nucho, Ayşe Parla, Tatiana Rabinovich, Karen Rignall, Lisa Rofel, Michal Ran-Rubin, Jonah Rubin, Chris Sargent, Kim Shively, Neha Vora, and Emrah Yildiz. I'm grateful to Sara Salazar Hughes and colleagues at the University of Southern California for the feedback that I received at the "Cities of Globalization" conference, and to Sultan Doughan and Frances Tanzer for the insights gleaned through the "Living Climate Refuge" event at Clark University. The "Cities of the Arab World" conference at Michigan State University provided the privilege of collaboration with and mentoring from Najib Hourani and connection to Hatim El-Hibri, Claire Panetta, and Aseel Sawalha. Thank you to Noura Wahby for the intellectual engagement, guest lectures, and invitations to speak with AUC students. Sherine Hafez and Susan Slyomovics became mentors and friends while I was at Whittier College and shaped my thinking about the anthropology of the Middle East and the academic profession. My time at Whittier was made infinitely better by Julie Colling-Dogrul, Anne Kakaliouras, Becky Overmyer-Velazquez, Olivia Solis, and faculty in the Anthropology-Sociology Department, the Environmental Studies Program, and the Digital Humanities Program. Nisreen Melhis became a friend whom I will carry with me wherever I go. Thank you to Ernestine McHugh, who played a role in teaching me to love anthropology and has remained a friend for more than two decades.

At Johns Hopkins University Press, I am tremendously grateful to the editors for the Water and Society series, Wendy Jepson, Amber Wutich, and Alexandra Brewis, who have provided support and encouragement for my project. Robin Coleman, Juliana McCarthy, Jane Medrano, Kathryn Marguy, and freelance editor Merryl Sloane treated my manuscript with care and provided the best possible experience for a first-time author. Additionally, Heath Sledge worked her magic copyediting the first full draft of the text; Melanie Magidow did a fabulous job with the transliterations from Egyptian Arabic; and Kari Anne Guilbault provided much-needed support with citations. Special thanks to the three anonymous reviewers for their very helpful feedback and suggestions. All remaining shortcomings are my own.

In many ways, the seeds of this project were planted by my father, John Farmer, in our many discussions about water in eastern Washington during my childhood. After his passing, doing research on water felt like continuing a lifelong conversation. To my mother, Marcie Farmer; my sister, Elizabeth Farmer; and my niece, Layla

Rose, I owe a million thanks for providing the foundation on which everything else is built and for their unwavering support. Thank you to Sparky and Judy; my uncles Bill, Doug, Mike, and Tom; my aunts Amy, Beth, Dee, Janie, Joanie, and Nancy; and all of my cousins for years of encouragement, particularly through the toughest days of the pandemic. Thank you also to Hamad and his family; Egypt would not have been the same without them.

Thank you most of all to my amazing daughter, Maryam. You have been on the journey of this book for the entirety of your life. Every word here is a stitch of time with you.

Chapter One

Introduction

Connecting Well

There is a saying in Egypt: *Law nahr al-Nīl ba'a ṣālṣa, mish ḥa-kaffī kull al-kūsā illī fīki yā Maṣr*, which translates roughly to "If all of the Nile River became tomato sauce, it wouldn't be enough to cook all of the zucchini in Egypt." In Arabic, zucchini (*kūsā*) is another term for nepotism (*wāsṭa*). The joke is that, like cooked zucchini (a very common dish in Egypt), nepotism is so endemic that the Nile River, despite its profound length and depth, is not big enough to contain the problem. This joke hinges on the central cultural and bureaucratic role of connection, or the ability to get what you want through networks of relations in ways that exceed, circumvent, or grease the official channels for achieving things. In order to accomplish a range of everyday tasks (like securing an official water connection), Egyptians often rely on personal or family relationships that they have forged and carefully tended, sometimes called *wāsṭa*.

Being *well* connected is often thought of as an elite project in which social, political, and economic networks are cultivated by wealthy individuals to maintain their class advantage.[1] Such extralegal or interpersonal engagements with representatives of state institutions in the region are often classified as corruption, bribery, and nepotism. They are labeled by many academics and global institutions as indicators of bad governance.[2] The purported solution is to eliminate such practices so that government functions can be carried out in transparent, accountable, and predictable ways. The salty tone of the joke displays the frustration that Egyptians feel at the necessity of operating through networks, at living in a world in which both advancement and survival are unevenly possible for those with weaker or less connected networks.

While the *wāsṭa* joke focuses on nepotism—points of connection to official domains—those who live on the urban margins deploy connections in other realms equally fundamental to life in Egypt. The life chances of these marginalized urban inhabitants are in many ways determined by the density of what could loosely be termed lateral and vertical overlapping social entanglements. Everyday Egyptians utilize the idiom of connection in generative ways and as a field of maneuver and improvisation. Being well connected in this sense is about mutuality, solidarity, connections of preference and community, and ties of affection and irritation. All points of contact have some friction (Tsing 2005), and in Egypt, this friction is lubricated by social elbow grease.

In this book, I follow residents of the informal urban area of Ezbet Khairallah as they labor to exchange, value, and evaluate potable water and to eliminate wastewater in their area. The channels and junctions of social webs help to shape the flow and distribution of water, particularly for those living in the dense informal neighborhoods of Egypt's capital city, Cairo. Water's flow and stoppage are mediated by social, material, and moral channels that urban residents labor to sow, locate, hedge, and sense their way through. Their practices are a locally specific form of the sociality of water, and they are integral to the project of being well connected—to other people, to their neighborhood, and (quite literally) to official water systems via pipes, plumbing, and faucets.

This volume is the end point of an investigation into how people create connection, an active project that is always in the making, and how they connect well. Residents labor to make "well enough" or "good enough" feasible in circumstances where connections are both necessary and, at times, double-edged. While the cultural form of connecting well in Egypt is broad and deep, here I consider it in one particular context: how it operates in projects related to water and wastewater in an informal urban space. The term "well connected" describes the various projects, embodied practices, and habits of social relations that quench people's thirst, help them clean their home, and help them rid themselves of sewage in a particular place and time: Ezbet Khairallah (also called Ezba) in the early twenty-first century.[3]

Let me lead you to water in Ezba, though wait before you drink. In 2009, when I entered Ezba, I first encountered water in the sewage puddles gathering at the bottom of a limestone hill. This hill stretched up to a plateau covered with the common redbrick of informally built (unregistered or unapproved) housing in

Cairo. Before I could see the wastewater, though, I smelled it. The sewage, which had escaped the confines of septic tanks, bubbled up in pools. It changed the form and shape of the road. Where the wastewater had dried in the hot sun, it tightened the road into deep, compacted waves of baked mud. As I moved into the neighborhood, off to one side I saw a grouping of three small earthenware jars called *ʿullal*. This was a *sabīl*, which had been created to offer drinking water to anyone passing by. *Sabīls* offer water in the spirit of connection and mutuality with those living in and passing through the area—an emblem of a larger moral economy of gifted water. On a hot June day, the water in the *ʿulal* would have been cool to the touch and have the characteristic taste of warm brick.

Just past the *sabīl* was a bathtub full of live catfish, barely covered with water, near a fish stand where their predeceased brethren were lined up on newspaper awaiting customers. A middle-aged woman was picking through zucchini at a small vegetable stall, perhaps to make *maḥshī*—the stuffed zucchini cooked in tomato sauce referenced in the joke—for a family meal. The afternoon sun drew out sweat, which poured from the bodies of those exiting microbuses coming in from other areas near downtown Cairo. One young woman wiped the sweat from her brow with a tissue and dropped it on the side of the road. Someone else took a sip from a plastic bottle of water, carried from home or filled from a *sabīl*, which was wrapped in a plastic bag to contain the condensation. A young man with a hose was spraying down the dirt in front of the shop where he worked, trying to contain the dust from the unpaved roads, as the neighboring shopkeeper leaned against her counter and gave advice on the amount of water he was using. It should be just enough to tamp down the dirt, she said, but not so much that it filled up the septic tank at the doorstep of her building. All water that appears on the streets of Ezbet Khairallah was open to contestation.

In this book I attempt to answer some basic questions about how residents of informal urban areas obtain and manage that irreplaceable substance so critical to human life: water. I closely examine how water comes to matter to people in an urban neighborhood, including how material systems, political scales, personal relationships, and moral frameworks intersect to shape the distribution and movement of water. I look at how wastewater and drinking water facilitate the construction and negotiation of neighborly relations to make a place feel like home. By tracing the labor invested in being well connected and the everyday practices through which connection is enacted, I offer insight into how those living on the urban margins construct livable lives—indeed, how they flourish—in places often structured to marginalize them. I follow the winding and ramified implications of improved water and wastewater infrastructures by documenting the installation of a state-funded wastewater system. This discussion adds nuance to the goals of universal water access and improved sanitation, which are so central to public health and global development. I show not only the driving need for these public goods but also the kinds of sociality and relationships that the gush of universal, on-demand water might sweep away.

Well Connected Is Material

From the cement sewage pit to the long line of women holding pots to collect water, connecting well is made material in water itself. A simple reflection on water's materiality calls to mind its heaviness, its capacity to dissolve a wide variety of particles, its response to pressure, its boiling and freezing points, and its unique role in the constitution of all biological life. Veronica Strang (2004) has detailed the ways in which our interactions with water teach us about being human and how water intersects with the ways in which people construct ideas of what home is like. Science and technology studies (STS) have meaningfully expanded scholars' focus of attention beyond the human to more deeply engage with both biological and nonbiological others that make human life possible and that also simply exist meaningfully in the liquid world (Lyons 2018; Neimanis 2017; Hastrup and Hastrup 2015; Linton and Budds 2014; Helmreich 2011, 2009). Understanding water as H_2O, a unitary and ahistorical substance, is a relatively recent development in Western thought, a departure from earlier interpretations of water as multiple and informed by context (Linton 2010). Water must be understood as a material substance that is deeply situated, informed by social and material relations in particular times and places.

As a material substance, our body needs water to survive. But not just water in the abstract. We need the right quality of water at the right times and in the right quantities. We need to distance ourselves from it once it becomes a flood or a locus for parasites, bacteria, and other pollutants. We need to bring it closer to drink, to bathe, and to make homes, lives, and neighborhoods. Therefore, in each place and time people come up with mechanisms for managing their requirements for water, requirements that hinge on temporality, quality, and quantity. In this book I offer a broad engagement with water in multiple aspects and a close attention to its particularities through a prolonged engagement with a single neighborhood. I build on earlier scholarship that is attentive to local somatic knowledge of water in Cairo (Ismail 2018). I center interpersonal relations as key to the circulation and management of water, thereby decentering the state-society relations more commonly discussed in scholarship on water infrastructure.

The materiality of water poses tensions in my exploration of human-water relations. Scholarship about water must preserve the qualities of its materiality while also tracing how social worlds encounter water—a substance that is always already shaped by human endeavors and explanatory frameworks. STS scholars remind us that the sociotechnical systems that manage water—from urban potable water networks to large-scale dams—shape when, how, and where people encounter water (Swyngedouw 2015; Rademacher 2011; Gandy 2004, 2003). The social and the mechanical are intertwined in these material water networks, for as Larkin (2013) reminds us, infrastructure is oriented to addressees, who both read and are made through their relations with these systems. Pipelines, pumps, and joints are a material scaffold to which social and political worlds are anchored; they are material surroundings that emerge from and are layered over with attachments, habits, and diachronic imaginations. And water itself is what connects these things.

Infrastructures also touch on questions of governance and the undergirding frameworks for the constitution of life in the modern world. Tracing the materiality of infrastructural networks at various scales can make visible the nature of states, their practices of governance, and layers of interaction between representatives of state or state-like entities and the people variously labeled as consumers, customers, or citizens (Stamatopoulou-Robbins 2019; Anand, Gupta, and Appel 2018; Larkin 2013; Star 1999). As Nikhil Anand has noted in his influential *Hydraulic City: Water and the Infrastructures of Citizenship in Mumbai* (2017), the particularities of water infrastructure both help to constitute the political mechanisms of a liberal city and produce relations that are other than liberal. Anand argues that the particular forms of Mumbai's water systems make possible the "right" kind of function of municipal systems and at the same time create the necessity for shadow infrastructures—ways for illiberal, informal systems to get water.[4]

To understand the many ways in which people, particularly marginalized people, actually operate to make life possible, we must understand the limits of liberal conceptions of the state, which posit that certain functions, from extending utilities to adjudicating land tenure questions, can only be correctly carried out by formal state institutions. Under this conception, alternative or illiberal practices (such as the patronage networks that marginalized communities often use to ensure services) are read as signs of corruption or insufficiency, as signifying a failure of the state (Wahby 2021; Sultana and Loftus 2019; Khalil 2018). A great deal of development work aims to eradicate these illiberal practices under the rubric of institution building (Adly 2020; Stamatopoulou-Robbins 2019; Bayat 2013), and the presence of these practices is often used to denigrate postcolonial states as having weak or bad governance—a modern version of a long-standing imperial discourse about the inability of racialized others to self-govern (Hafez 2013). To understand these informal networks of exchange and circulation, we must broaden our conceptions of the state and how we define the success or failure of governance.[5]

Water infrastructure is not only about formal and informal networks; it is also about modulating the separation between types and states of water. Scholarly research on water tends to follow one network for managing one iteration of water in the city at a time: potable water, wastewater, urban runoff. But in urban spaces, as elsewhere, potable water and wastewater intersect. In this book, I examine potable water and wastewater together (including overflow and runoff, though in arid Cairo, urban rainwater runoff is not very significant), showing how the mechanisms and gaps of one set of systems can create challenges and tensions in another. During my time in Ezbet Khairallah, the mutuality and interdependence of water and wastewater were particularly visible. For example, the long-term use of septic tanks imposed volume limits on the amount of potable water that could be used, even for people with (unregistered) connections to potable water networks. Additionally, the nature of the self-installed potable water networks often meant that pipes cracked or had pervious joints, allowing for contamination from the overabundant sewage water.

Well Connected Is Social and Moral

Water is a "total social fact" (Orlove and Caton 2010, building on Mauss 1990), an object that intersects with all levels of human endeavor. Because it is so polyvalent, water makes a useful focus for social science scholars' attention. Andrea Ballestero (2019c) has called water a "learning machine," through which anthropology as a discipline has come to understand a great deal about how particular kinds of social relations are organized, about how people think of their material world, about what kinds of moral codes are enacted in the ways that people interact, and how those might change over time.

As with its materiality, water is not simply "social" in general, but rather must be contextualized in specific societies and cultural worlds at particular times and places because those illuminate how and why water appears where and as it does. Importantly, such social universes cannot be assumed to be coherent and unilinear; they contain contradictions and are subject to multiple interpretive frameworks (Gouda 2016; Wagner 2015; Rademacher 2011; Coles and Wallace 2005). One common set of binaries that the anthropology of water attempts to challenge is scientific knowledge (knowing H_2O) versus traditional systems of understanding and engaging with water (e.g., baptism). Such binary interpretations reproduce a narrative of the linear progress of secular knowledge arising as a movement away from superstition and irrationality, and they position other ways of knowing and engaging with water as always somehow suspect (Bagir 2005; see also Alley 2002).

Social relations around water always involve questions of power and, crucially for this book, gesture to the role of water as a medium through which political and social stakes are negotiated. Particularly at the level of local and everyday forms of well-connectedness, water is tied to moral frameworks and the ethical systems that enable their enactment. Moral and ethical systems are central to organizing water governance, where they function to assign and mediate the worth of subjects, objects, and actions (Groenfeldt and Schmidt 2013). Water is implicated in the moral geography of cities, as city spaces both articulate and reproduce ordering logics through unequal access to resources (Schafran, Smith, and Hall 2020a, 2020b; Peteet 2017; Zeiderman 2016; Harvey 2008; Gandy 2004; Smith 2000). And water poses many challenges to cities, including threats of inundation from river flooding, sea level rise, and extreme rainfall; insufficient or aging infrastructure; threats of paucity from drought, heat waves, and pollution; scarcity in the large rural water catchment systems that make city life possible; and inequitable distribution.

In Ezbet Khairallah, as elsewhere, questions around water are moral questions. Morality is present in several dimensions in the discussion of water, and I highlight three here: religious systems, networks of reciprocity and social norms, and the moral tenor of shame.

The most easily identifiable of these are the religious frameworks. The dominant religions in Egypt are Sunni Islam (roughly 85 percent of the population) and Coptic Christianity (roughly 15 percent of the population). Ezbet Khairallah is a mixed religious neighborhood (as are most areas in Cairo) with several mosques and the important church of Tahuna. In Islam, water is considered a gift from God, and the doctrinal frameworks for the management, sharing, and use of water are important guiding principles. The significance of water in Islamic thought is clear in the often-quoted Qur'anic verse "And God has created every animal from water: of them there are some that creep on their bellies; some that walk on two legs; and some that walk on four. God creates what He wills; for verily God has power over all things" (Qur'an 24:45). As part of the Eastern Christian religious tradition, Coptic Christianity also recognizes water as divinely provided and as a mechanism through which purification happens (Denysenko 2012). These religious ideas arise in the state's framing of unregistered water use as theft, in moral understandings about the importance of household-level sharing, and in the context of the charitable gifting of water in the form of *sabīls* (Wutich et al. 2018; ElDidi and Corbera 2017; Schmidt 2017).

The ethics of water are also expressed through social norms, particularly those around hospitality, cleanliness, and interpersonal obligations (Meneley 1999). How sophisticated are people at managing their (often water-mediated) relationships with neighbors? How industrious are women in maintaining the cleanliness of their homes and the health of their families? How devoted are people to the obligations implied by having sufficient water in a context of scarcity? Household water sharing is organized by cultural and religious beliefs about how, when, and with whom one shares an expectation of mutual obligation (Wutich et al. 2018; Naguib 2015).[6]

The third key dimension of water ethics is shame (which is, of course, tied to the social in many ways). Here, I examine shame in terms of wastewater management, which is a socially inflected moral issue because poorly managed wastewater affects how others see Ezbet Khairallah and the moral character of its inhabitants. Ezbet Khairallah is a large informal area in Cairo, and its residents are primarily lower-income migrants from Upper Egypt. Over a period of forty years, residents collaborated to build their own utility infrastructure, creating illegal connections to state water mains in adjoining neighborhoods and installing basic septic tanks (*transhāt*) in the roads in front of homes. The neighborhood's material difficulties compounded the marginalization of the community, which was already understood as a lawless and dangerous place harboring drugs and prostitution.

In this study of water, I discuss the responses as Ezbet Khairallah received a state-sponsored sewage system in 2008–2010 and a piped potable water system in 2011–2013. This allows me to get at important questions about how people apply ethical frameworks in their social relationships: what it means to be a good person, how someone estimates their own morality (Deeb 2015; Hafez 2011), and how other people interpret them as residents of an informal urban neighborhood. In this book, I incorporate the taste, smell, and feel of water quality, bringing together the material and the social aspects of being well connected.

Water in Egypt

At the national level, access to sufficient water is a pressing issue that has been at the center of some of Egypt's key political contestations over the twentieth and twenty-first centuries, from distribution of the Nile's waters, to agricultural policy, to thirst as a constitutive element of citizens' discontent. In Egypt, almost all surface water that can be made consumable originates outside the nation's borders. The Greek historian Herodotus famously said that Egypt is the gift of the Nile. The Nile is generally regarded as the longest river in the world, flowing some 4,000 miles.[7] It connects Egypt to its upstream riparian neighbors in fraught, long-standing relationships. A river of ink has been spilled about the importance of the Nile to the changing geographical and social forms of Egypt over time and about the relations of force and precedent (and the legal and material arrangements for the allocation of water they produce) along the Nile's length (see Derr 2019; Barnes 2014; Mitchell 2002). It is likely true that there is no Egypt without the Nile, as some 95 percent of the population of 100 million people live in only 4 percent of the total land mass of Egypt: in the fertile valley of the Nile (Abd Ellah 2020). The remaining 96 percent of the nation is classified as nonarable land, primarily desert, and is sparsely populated (Abd Ellah 2020).

At least, this is what the development sources tell us. National snapshots and the accompanying international testimonials about the dire state of the world's water supply, however, do not unpack the science behind such measurements. What hidden guesses, hunches, and assumptions lie behind the framing of this picture? The image cannot contain how Egypt's borders and middles blur and transgress the official maps that calculate its area in square kilometers. Look, these stories say, at how narrow a band of habitable and fertile land there is in Egypt! Look at the map! It shows the slender ribbon of the Nile River slipping through a rectangle of sand. How little room to grow, the story says—to grow Egypt's population and to grow the crops to feed them. While there are real material concerns about water availability in Egypt, the framing of this story depicts the country as in need of expert management, as a picture of inequality that requires development—from international organizations, such as the World Bank and the International Monetary Fund (IMF), and to some extent from the governing structures of the country itself.[8] Crisis narratives can be both productive and dangerous because of the urgency that they imply. This urgency can enable more active engagement by the broader public or a more authoritarian control of resources. Crisis narratives can authorize "green grabbing," in which governments or international entities push to take more extensive control over resources or space to ensure humanity's "common well-being"—often at the expense of marginalized populations (Rignall 2019; Fairhead, Leach, and Scoones 2012). How do we engage with material challenges in ways that do not authorize additional violence against the most marginalized? This is a question with its own urgency, but my goal here is much more modest: to understand the effects of the broader water context in Egypt on municipal water systems.

According to international measures, Egypt is a water-poor country. As of 2011, Egypt was at 70 percent of the water poverty threshold with 700 cubic meters/capita/year; the established scarcity threshold is 1,000 cubic meters/capita/year. As of 2022,the Egyptian Ministry of Water Resources and Irrigation adjusted that figure down to 570 cubic meters/capita/year, and it is expected to drop below the absolute scarcity level (500 cubic meters/capita/year) by 2030 (FAO 2016). As with all stories of calculation, the numbers hide vast disparities in the allocation and distribution of water resources within the country and between various users, with small farms, rural areas, and poorer urban neighborhoods suffering the most water stress (Albuquerque 2010). In real material terms, the present water system does not have sufficient inputs to maintain the current usage patterns, and the trend is growing worse.

Currently, agriculture swallows up the vast majority—85 percent—of the water extracted from the Nile in Egypt (Jagannathan, Mohamed, and Kremer 2009). There have been a number of large-scale projects aimed at building infrastructure to move water both for agricultural purposes and as part of the national project of reclaiming desert land. In addition to Lake Nasser and the Bahr Yussef Canal, which feeds the oasis of Fayoum and dates back to pharaonic times, there are several other large water diversion projects, including Toshka off Lake Nasser (Barnes 2014) and several projects in the Sinai. Jessica Barnes's (2014) work on larger water systems in Egypt is instructive in its attention to the daily processes of resource production. As Barnes demonstrates, agricultural water that reaches end users is produced by systems that enable or deny supply in particular places at particular times.

While the Nile is the main source of water for most agricultural, industrial, and domestic uses, soil water and groundwater are part of Egypt's future-tense story of water possibilities (Allan 2001). Egypt has four aquifers of substantial size. The Nubian Sandstone Aquifer is by far the largest. It is unique both in its size, covering some 2 million square kilometers across Egypt, Sudan, Libya, and Chad, and for the fact that it holds fossil water. Fossil water has been trapped in an underground location for a very long time, possibly thousands of years, and it does not receive substantial recharge from precipitation.[9] The other three aquifers—the Nile, the Moghra, and the Coastal—are meteoric water, which comes from precipitation, and are therefore (at least theoretically) renewable (Sayed et al. 2020; Hefny, Farid, and Hussein 1992). Many have proposed that some future water needs in the area could be met by tapping into these aquifers, although current withdrawal rates exceed recharge rates, and there is debate about how to best estimate an aquifer's yield (Pierce et al. 2013; see also Ballestero 2019b).

The talk of tapping these aquifers responds both to an increased demand for water and to new pressures on the water supply.[10] These pressures come from two main sources: climate change and the building of the Grand Ethiopian Renaissance Dam (GERD). Climate change–related rising temperatures and shifting weather patterns are affecting rainfall in upstream riparian countries, increasing evaporation throughout the water system, reducing aquifer recharge, and instigating sea level rise, which pushes salt water into coastal aquifers. Climate change puts particular stress on Egyptian agriculture, and the United Nations World Food Programme estimates that southern Egypt could lose 30 percent of its food production capacity by 2040. The overall amount of water available is decreasing, and the allocation of the remaining and uncertain water is also at a moment of crucial flux.

The building and filling of GERD is upsetting long-standing (and long-contested) allocations of Nile waters, which have privileged Egypt and, to a lesser extent, Sudan over other Nile River nations. The 1959 treaty between Egypt and Sudan allocated all of the water of the Nile River to the two countries, with Egypt receiving the lion's share at 55.5 billion cubic meters and Sudan receiving 18.5 billion cubic meters of water annually. The rest of the water was supposed to flow to the Mediterranean Sea; none was allocated to nine upstream neighboring countries (Rwanda, Burundi, Democratic Republic of the Congo, Tanzania, Kenya, Uganda, Ethiopia, Eritrea, and South Sudan). Egypt had long worked to maintain this privileged allocation by suppressing political dissent, but this has proved increasingly difficult in the twenty-first century.

In 1999, the Nile Basin Initiative was established in an attempt to increase cooperation among the riparian neighbors (Allan 2001). Then, Ethiopia announced plans to build a large-scale hydroelectric dam on the Blue Nile River, which provides more than 80 percent of the downstream flow. Construction of GERD began in April 2011, and the filling of the dam began in July 2020. Egypt vehemently contested Ethiopia's right to build the dam. When this was unsuccessful, it sought to force a legal agreement between the two countries about the construction and filling of the dam, stipulating the allocation of water that Egypt would receive during times of drought. Egypt sees GERD as potentially decimating its farming and reducing its own hydroelectric production. In July 2020, President Abdel Fattah el-Sisi committed to continued negotiations with Ethiopia but also left open the possibility of military action: "Our right is not just to water but to life. No one dares encroach on us, on our interests or our national security" (Hendawi 2020). GERD has proven to be a politically sensitive topic for regimes in Egypt since its inception in part because it has challenged Egypt's projection of power.

Internally, the el-Sisi government has sought to lower demand in anticipation of reduced supply. The Ministry of Water Resources and Irrigation has limited the planting of some water-intensive crops and urged farmers to switch to seeds with shorter life spans that require less water; it has proposed the use of more efficient irrigation systems and trying to recycle more water. Egypt is also seeking to reduce household consumption. Prime Minister Mostafa Madbouly has ordered the installation of water-saving tap devices in housing, called for the building of more desalination water plants and wastewater treatment plants, and urged the government to impose harsher penalties for the theft or waste of public water (Samir 2020).

One of my contacts in Ezbet Khairallah told me in 2018 that rate hikes for municipal water were in fact a plan by the state to get Egyptians used to using less water in preparation for a diminished supply post-GERD. While the rate hikes had more to do with structural adjustment policies in Egypt, particularly after it signed a multibillion-dollar loan agreement with the IMF, this belief nevertheless points to the way in which municipal water consumers read the (un)trustworthiness of water governance in general.

Managing Urban Water

In the nineteenth century, as cities expanded in size and density, the need for water management expanded. Urban water management became professionalized over the nineteenth and twentieth centuries, and the tension that emerged out of the Industrial Revolution—between public and private water management, between local and municipal authorities, between networks constructed by for-profit enterprises and those created and maintained by the state—has only intensified. The question of who manages urban water is a complicated one, with answers ranging from local independent water collectives to large water networks.

Since the late twentieth century, the governing question of all these smaller debates about urban water has crystallized around a single question: who (or what) can best ensure universal access to potable water, the safe removal of wastewater, and the responsible use of limited resources? But the answers to this question assume that the goal is Western-style water and sanitation infrastructures. "The state" writ large is assumed to be at the very least the arbiter and guarantor (if not the executor) of water systems for citizens, and debate rages around the efficacy and efficiency of public, market, and nongovernmental entities (Schmidt 2017; Bakker 2010). This is important because the state—as a gloss for the collective governance structures that formulate, enact, mediate, and enforce operational configurations—is a powerful contemporary framework for thinking about how water management operates and how the management of water shapes the constitution of governance itself (Harris, Goldin, and Sneddon 2013; Adida and Girod 2011). At a very basic level in Ezbet Khairallah, expectations wavered. In our conversations, many residents expressed that the state *ought* to make better, more plentiful water available, but they also felt it probably wouldn't do so without being forced or cajoled—activities that require some form of connection.

State entities earn legitimacy and exercise control over spaces and populations through their management of water, moving it from rural to urban places, managing the volume and pressure of flows in rivers and pipes, cleaning it for consumption, and organizing its distribution across space and over time. This is what Anand (2017) calls the "hydraulic state," in which the state itself is in part constituted by and through water, and (hydraulic) citizenship (read as one's recognized membership in a state) is in part mediated by water.[11] In an important study of informal urban water practices in Cairo, Noura Wahby (2018) documents local patronage geometries and claims-making processes by both poor and wealthy urban residents within the paradigm of state-society relations.

But in many (perhaps all) places of the world, access to water is not universal. Gaps in water coverage prompt the usual story of technological progress, in which modern water and wastewater systems must be produced through the right combination of political will and money. Certainly, improvements to water and sanitation infrastructure have served to improve the health and quality of life for many around the globe. But this political, social, and technological story is complicated. Rather than simply needing to figure out how to unfold this technology faster, cheaper, and more enduringly around the globe in order to give everyone equal access and ensure the efficient use of finite water supplies, we need a fuller understanding of how necessary domestic labor and local social customs are to enabling these technological systems of water and wastewater management. The installation of formal water and wastewater systems will not necessarily fully solve the problems of access to potable water and creating a safe distance from waste. And as we see in Ezbet Khairallah, these formal systems—even those desired and fought for by residents—have often-ignored implications for human well-being and the advancement of standards of living.

In terms of environmental justice, which is an action-oriented framing of moral concerns, addressing water equity extends beyond just who has access, and includes determining who has to do the work to make it happen and what it costs in time, money, labor, and opportunity (i.e., the option to do other things with those resources). This has gendered implications, as the ways in which water is managed play an important role in the constitution of gender roles, social inequalities, and divisions of labor (Sultana 2009; Crow and Sultana 2002). Sociologists Ribot and Peluso (2009) have argued that access should be understood not as the "right" to derive benefits from resources, but instead as the "ability" to do so. This approach better accounts for the multiplicity of ways that people are able to access resources that extend beyond the logic of property. Accessing potable water in particular is often a task socially assigned to women, particularly those who are poor, impacting their livelihoods and lives (Sultana 2011). Projects for ecological justice must take into account moral and social ecologies among neighbors (human and nonhuman) and between residents and their neighborhood (Rubaii, forthcoming; Scaramelli 2021).

Over the course of my research, it became clear that the assumption that national governing structures are *the* producers and regulators of potable water does not always attend to the social relationships that organize water. Indeed, it rests on an outdated notion of "the state." Scholarship since the 1990s has seen some dissolution of the theoretical conception of the state, calling into question the idea of the state as a monolith and instead viewing it as being constituted through everyday practices. Twenty-first-century scholarship engages with the state-like effects of nonstate entities and examines the hollowing out of state-based sovereignty.[12] Social studies of infrastructure have also changed the terrain of attention to local and municipal governing structures in important ways. Yet anthropology has largely retained a liberal disposition toward the state as the appropriate mechanism through which people's needs are met. Although theoretical critiques of the state are now common, they are often predicated on an underlying desire for the state to be better at "stating," to fairly and efficiently provide services.

As the theoretical continuity of the state has dissipated, the ideas of citizenship and governance at other scales have also changed. But the liberal idea of the state as the central actor has retained its hold on our imagination in ways that make it difficult to see the sociality of water except through the framework of state-society relations. In the process of disaggregating the state, state-society relations have become so diffuse that they seem to have become nearly the exclusive paradigm through which we see the social. While governance structures of one sort or another often set the parameters for many of the conditions of possibility that people are trying to work within, people interact in ways that exceed their relations to systems of authority.[13] Being well connected is not just about proximity to government pipes, but also about the continual construction, maintenance, and management of local social networks.

I do not seek to discard the state nor to reinstitute the triumvirate of state–civil society–population, but instead to dilate the eye, to reattend to other forms of activities, other patterns of relation that are necessary to human well-being, for they affect the management and obtainment of resources such as water. As I show, Ezbet Khairallah epitomizes this dynamic: residents work to get water not only through the state but also through social networks and neighborly relations. By centering well-connectedness, in this book I make visible the multiple frameworks for making water work, which include but are not limited to stateness. The sociality of water encompasses the unfolding of human well-being through the provision and management of water, but this water is often provided and managed through love, social contacts, moral ideologies, interpersonal relationships, friendships, domestic rhythms, embodied sensibilities, and the everyday labor of connecting.

As ethnographers, we become complicit in power structures when we describe the world primarily in the terms that come with power—citizenship, state, society—even if our intent is to disrupt. Here, I begin from the cup, the puddle, the doorstep, and then look up to the larger forms and structures these suggest. What I saw during my fieldwork was a set of patterns and locations that cut across the standard categories that organize our scholarly thinking. By following the material in order to understand the social, I attempted to step outside the highly theorized paradigms of water governance and see what people were actually doing. By asking people about materiality, I found the social, material, and moral relationships that were dialectically constituted with the presence, accumulation, movement, and distribution of water in Ezbet Khairallah. Over the course of my research, my twenty-first-century anthropological framing of the issue changed. I no longer approached it as studying how people get water in nonnormative (i.e., extrastate) ways. Instead, I found myself asking about how water comes in and goes out, about the material systems that guide or don't guide its flows. Eventually, I began to see the ethical and social systems that shape these water networks, and the differences in how local residents and visiting anthropologists describe these urban waterscapes.

Potable Water Infrastructure

Modern water infrastructure in Egypt has reflected the changing organizational imperatives of institutional regimes as they shaped the material and social worlds to their own ends. British colonial projects aimed to configure water infrastructure in ways that supported the extraction of resources (such as improving agricultural water systems to enable the production of cotton and sugar) or that facilitated the management of labor (as with the water diversion projects set up to feed Ismailia and the workers building and maintaining the Suez Canal) (Gamal-Eldin 2020; Derr 2019; Mitchell 2002). Colonial governments improved water systems in urban centers like Cairo, their seat of power, instituting potable water and wastewater piping systems. In Cairo and Alexandria, water and wastewater infrastructure were developed further as the imperial government connected the cities to other locations.[14]

In the couple of decades following the 1952 revolution that ended the monarchy in Egypt, technological advances such as the Aswan High Dam operated as a testimony to the viability and prowess of the new nation-state. International development structures had emerged in response to the humanitarian and financial turmoil of the post–Second World War years (Sellman 2018; Mitchell 2002). During this period, states' welfare responsibilities were extended to both their own citizens and those of other states; in a sort of humanitarian imperialism, nations took on the duty of policing other states' treatment of their own citizens. The League of Nations (which became the United Nations), the World Bank, the IMF, and other postwar organizations sought to build the capacity of "weak" or "undeveloped" states to provide public goods, such as roads, schools, potable water, and wastewater management. Under this paradigm, "developing" states were described as needing financial, technical, and managerial assistance from "developed" nations to achieve these infrastructural goods.

With Egypt's Infitah (open door) economic policy instituted by President Anwar Sadat in the early 1970s, water and other infrastructures were part of the great promise of the postcolonial nation-state, touted as development for the good of the population. Farha Ghannam (2002) has discussed the discourses of modernity and civilization that were utilized by Sadat in his attempts to remake the city of Cairo into an urban metropolis that would draw tourists and international capital investment. Through new spatial divisions, Sadat aimed to revitalize the urban sphere while at the same time normalizing and disciplining poor people into becoming productive members of the nation. However, the real political lessons of such areas were more about the costs of being recognized by state entities, where being acknowledged means to be constituted as a problem to be solved. The flip side of this "cunning of recognition" (Povinelli 2002) was the disconnect between state policies and the interests and actions of bureaucrats interacting with the public; this gap was filled with the petty bribes that bureaucrats extorted from poor residents to look the other way on the illegal but ubiquitous small-scale, informal building projects.

Since the 1990s, development in the formal housing sector has been primarily aimed at wealthier segments of the population, with significant government investment and subsidization of infrastructure for expansion into the desert. Informal expansion, until this period mostly ignored, was officially frowned upon, and in 1996 it was criminalized: harsh punishments through military courts could be levied for any new building on agricultural land or for urban construction without appropriate permits (Sims 2012: 68). The fear of substantial repercussions created a short lull in the expansion of informal areas, but, according to Sims, the system soon righted itself, as government officials simply requested higher bribes for looking the other way. Despite these laws, the building of informal housing continued at a rapid pace, propelled in part by the increase in migration to the Gulf countries for work in the oil fields and in the construction sector. These jobs provided the necessary capital for many more lower-income Egyptians to build foundations for multistory family buildings. Compared to the investment in desert satellite cities, relatively little government attention was paid to the significant needs of informal areas or to the outdated and insufficient infrastructure in older areas of Cairo (Davis 2006: 85). President Hosni Mubarak's increasingly "liberalized" economic policy instead prioritized economic development through land speculation and real-estate development (Khalil 2019; Wahby 2018). Such neoliberal practices have continued under President el-Sisi, and they sometimes take new forms, such as regime-security urbanization (Flahive 2018), in which the urban form is modified to deter mass public mobilization, including the relocation of important government ministries to the remote New Administrative Capital.

In the urban margins, the goal of a dignified and healthy life drives the unceasing labor to be well connected. While the kind of connection, or *wāsṭa*, that functions as a barrier to marginalized Egyptians trying to access governance structures and utility networks frustrates and challenges people in everyday life, connection is also a field of improvisation and humor. People use the idiom of connection to craft dense networks of solidarities and vexations and to curry favor—but not necessarily in the same way, with the same targets, or for the same ends. The ongoing work to connect well enables residents to sow and maintain connections to their neighborhood as they face repeated attempts to dislocate them. Residents deploy smart social labor to locate sources of help in their broader webs of relation and to bend multiple networks into meeting at the doorsteps of their homes (internal plumbing and external municipal systems, septic tanks and neighbors). They hedge connection as they try to manage the double-edged nature of networks, which can buoy and also possibly drown them. And residents expend the mental and physical labor of embodied discernment as they manage the threshold between potable water and wastewater.

In all of these ways, residents of Ezbet Khairallah seek to live well, to work through the challenges of their context to enact a life that meets their own locally inflected expectations of what that life course should look like. A key piece of living well in informal settlements in Egypt, as elsewhere, is the opportunity to drink healthy potable water and reside in dignified separation from wastewater. In this book I bring together the conditions for water's presence and absence in Ezbet Khairallah, its rhythms and temporality, its flavors and scents, its affective and moral registers in order to better understand how water shapes life possibilities and how, in turn, these interrelated systems shape social worlds.

Methods

As is often the case in ethnographic work, I arrived at Ezbet Khairallah by a somewhat circuitous route. I had originally intended to work in the western desert oasis of Siwa, researching how water helps the Egyptian state to constitute its national borders through extractive and developmental projects. My interest in water sprang from an earlier research project in the Siwa oasis on an ecolodge that used a narrative tying the local indigenous culture to a pristine environment as a consumable (and very high-end) package for the global elite. In that earlier project, the issue of water emerged as central to understanding large-scale changes ongoing in the oasis. After preliminary research in Siwa in 2007, however, it became clear this was not the right project for me.

I decided to look instead for a location in an urban center where I could focus on potable water issues as they might illuminate social and material structures in the twenty-first-century Egyptian state. A friend in the Ministry of Antiquities pointed to Ezbet Khairallah as an ideal location because the water from that neighborhood had become a major problem for the artifacts in the nearby site of Fustat (Abdel Adheem 2010; Obeydo and Maged 2010). During my visit to Ezba, it became clear that its potable water and wastewater systems were undergoing significant changes due to an ongoing installation of a municipal wastewater network, which would allow me to get a closeup view of large-scale utility structures in transformation. In 2009, Cairo was a hotspot for research in the Middle East, and the anthropology of water and infrastructure hadn't yet reached its current ascendance. Because Ezbet Khairallah was informal and had a reputation for danger, I couldn't get police permission to rent an apartment in the neighborhood. This turned out to be serendipitous. I settled a couple of metro stations away and traveled into Ezbet Khairallah every day, and the movement into and out of the neighborhood offered its own set of lessons about how tenuously the area was integrated into the city's material and legal networks.

My ethnographic research in Ezbet Khairallah focused on everyday life. I engaged primarily with residents of the area, spending days with women as they managed water-related tasks, such as doing laundry and dishes, filling water bottles and *sabīls*, and emptying septic tanks. I also interviewed people who worked in the water and wastewater sectors, people who worked in entities that were concerned with informal housing, and the area's political representatives. This book is based on nearly two years of fieldwork, the bulk of which was conducted over sixteen months between 2009 and 2011, during the implementation of the septic system. I was also there for two months in 2016 and three months in 2018, examining the increasingly clear long-term impacts of the Egyptian revolution (2011) and subsequent military coup (2013).

My methodology was developed to illuminate how water circulates materially and socially through the neighborhood of Ezbet Khairallah, how the processes of distribution and use of water and the management of wastewater are carried out within networks, and how the processes of negotiation between residents of Ezba and aspects of the utility institutions operate on the day-to-day level. I carried out research through semistructured interviews and participant observation in community homes and businesses in order to understand patterns of domestic consumption of potable water and the management of wastewater. The methodology of participant observation is particularly significant for this project's choice of the quotidian as the central analytical framework through which I have engaged questions about community and social networks.

During my first period of research, I worked with a key group of nineteen community contacts, which were divided among ten households. During weekly visits with each of these households, I participated in their daily tasks, particularly those that related to water. Many of my interlocutors were women and girls, given the gendered distribution of labor and my own female cisgender identity, but this is not a book about gendered water activities. The daily water-related domestic tasks included washing dishes, scrubbing rugs, doing laundry, cooking food, bathing children, cleaning the unpaved street outside the home and dampening it with water (to prevent dust from rising), feeding domestic animals, and collecting and storing drinking water in jars, jugs, and jerricans when the water was running. (The stored water tides over families during periods when the water provision is cut.) I developed long-term ties with the community—my own version of being well connected in Ezbet Khairallah—and I have continued to engage with them on return visits and over social media for almost fifteen years.

I also participated in tasks related to the establishment of the new sewage system in Ezba: watching the removal of septic tank contents, filling septic tanks with debris once they were no longer needed, providing assistance to contacts in Ezba who were working to repair damage done by sewage water to the foundations of their homes, replacing faulty piping, and remodeling homes after the sewage system was operational. The implementation of the sewage system in the area enabled me to further observe the day-to-day dynamics of water management, including negotiations between neighbors and between community members and representatives of the water sector. I also observed the unfolding of relations between the community members and the array of nonhuman (and sometimes noncorporeal) others with which they are connected, from parasites to jinn to the built environment.

Beyond my work with the community in Ezbet Khairallah, I interviewed representatives of government institutions and of local and international organizations that deal with water resources and sewage issues. These interviews were primarily conducted in Colloquial Egyptian Arabic, a language that I speak, and I have translated them into English. A handful of interviews with high-level representatives of international institutions and Egyptian agencies took place in English. The use of an interpreter was required for three interviews in Arabic that included highly technical language. Interviews were conducted in a variety of locations, as dictated by local conventions regarding private and public space and gender mixing. The names of the local interviewees from Ezba are pseudonyms; the governmental and NGO interviewees have given permission for their names to be used.[15]

One aspect of this research that surprised me on a number of occasions was that I was read as a partisan for Ezbet Khairallah residents. When designing the research project, I deliberately steered away from any clear association with formal entities, such as municipal authorities or nongovernmental organizations (NGOs), through which many anthropologists negotiate their first contact with a research community. This common method of connection is often based on a commitment to working with local civic organizations and avoiding exclusively extractive ethnography. While I share this goal of ethical fieldwork, I worried that in the informal neighborhood of Ezbet Khairallah, being enmeshed in NGO worlds would serve more to overdetermine residents' ideas about me than to offer opportunities to contribute to the community's well-being. To be sure, I was always read through the lens of my intersectional positionality, particularly as a female American researcher who fit (awkwardly) into my interlocutors' domestic social networks, and sometimes residents wished me to be a conduit to NGO systems. (They were, in essence, treating me as a particular kind of potential node in connecting well, as I detail in chapter 4.) But I did not want my work to be channeled through institutions; I was interested in the informal and local social negotiations around water. I think that this decision helped me analytically by decentering the state-society relations that predominate in the literature about infrastructure, allowing intracommunity and other-than-state relations to become more focused.

I planned to start with research in the neighborhood and then move to interviewing (not conducting ethnography with) governmental and NGO entities later in the projected sixteen-month period of original research. What my research plan failed to account for was how representatives of state institutions would read me differently when they learned that I had spent time in Ezbet Khairallah. My one-way imagination of social embeddedness came back to bite me on several occasions when I tried to follow up on the details of the official water systems in Ezba. For my first visit to the Ain el-Sira complaint center, I went through a connection in Ezbet Khairallah, trying to use the relations that my Ezba conduits afforded. Upon arrival at the complaint center, however, it became clear that I didn't fit the usual categories of visitors. I didn't belong with my Ezba contact, for I was not an ordinary resident of the areas (formal and informal) that the office served, but I didn't enter through official channels, which foreign researchers might conceivably do when contacting a minor utility office. Workers were confused, not knowing the right person for me to talk to or even which utility office was the right one.

As time went on, I learned to be savvier in my self-framing as I reached out to officials and NGO actors, cultivating the power of my institutional affiliations in the United States (University of Texas at Austin) and Egypt (the Fulbright Commission, Cairo University). On a few occasions, I called to make appointments with officials before completing all of my prep work, thinking that I would have to wait to be seen, but I found instead that doors opened quickly in response to those institutional affiliations in this postcolonial state. Calibrating connections to do them well, it turns out, was not only the object of this ethnography, but also its mechanism.

Organization of the Book

I examine the processual and multilayered nature of connecting well, offering an analysis of four modalities of being well connected in action: sowing, locating, hedging, and sensing. These four modes elucidate the connection between potable water and wastewater, trouble the generally positive ethical valence of water systems, illuminate embodied ways of discerning water quality, reveal crucial everyday practices that impinge on human health, and identify key junction points in water and wastewater systems that are mediated by neighborly relations. I also deepen attention to domestic spaces within infrastructural studies and trace connections between internal and external systems. Much previous literature about water has focused on state agents, NGO experts, and local actors, with less attention being paid to households and domestic spaces as key to the functioning of urban water systems. I examine these networks in the context of substantive systemic change, describing the process of implementation of the wastewater system and offering insight into systems in flux. Through attention to the complexity of implementing water and wastewater infrastructure, I offer an ethnographic view into local interactions with water systems that have major impacts on public health. In the conclusion, I examine the form of the *sabīl* (charitable water fountain) as it gathers connection as a way to extend the themes of the book and point toward future research.

Chapter 2, "Sowing Connection," examines the establishment of Ezbet Khairallah's foundations. Sowing happens as residents work to implant themselves materially, morally, legally, and socially in their location. Connecting well in this case is about making deep roots, firmly embedding residents in their area and enabling them to reach tendrils out to adjoining areas to draw in the necessities for people's persistence in Ezba. The story of Ezbet Khairallah demonstrates the deep ties between land tenure and utility access, and in this chapter I document the forces that fight to keep this implanting unsettled. Sowing also happens in the everyday, and I show how people work to make life possible on what started out as desert or "waste" land. Connecting well in this sense is about growing a lively and vibrant community on this land through everyday practices of obtaining, exchanging, and utilizing water in the neighborhood.

Chapter 3, "Locating Connection," builds on the account of the labor invested in making social relationships and material systems connect. I detail the everyday practices embedded in a specific kind of labor (managing connections between potable water and wastewater) carried out by a specific group of workers (primarily women) in a specific location (Ezbet Khairallah). The women mediate between the water and wastewater systems, both as they exist inside the house and as they change outside the house with the implementation of the municipal sewage system. I demonstrate how all of these material, administrative, moral, and social systems are brought into connection and are mediated at the doorsteps of homes. I also tell the story of one woman's septic tank, which lays out the moral implications of being enmeshed in sewage and demonstrates the neighborly labor that closed this septic tank when the new piped sewage system was implemented. In this chapter, I ground the story of the domestic as key to the functioning of water utility networks.

Chapter 4, "Hedging Connection," focuses on the usefulness of cultivating ambiguity in water connections in three situations: the relationships that representatives of state institutions cultivate with dwellers in informal areas of the city, to whom they offer minimal services; the relationships with state representatives that Ezba residents cultivate in response to the state surveillance that has accompanied the implementation of municipal utility services; and the relationships between Ezba residents themselves as they engage with each other in a context where social networks both promise help and can potentially overwhelm scarce resources.

Chapter 5, "Sensing Connection," investigates the multiple ways in which perspicacity and sensory experience are key to evaluating the "well" in well-connected water systems. In particular, I examine embodied forms of discernment, concentrating on smell and taste as key ways of assessing water quality. I also offer the larger context in which Ezbet Khairallah's water-quality struggles emerged, examining the history of water-quality incidents in contemporary Egypt. I lay out the deep relationship between potable water and wastewater—a relationship that shapes the challenges of ensuring an adequate supply and the appropriate quality of potable water.

The conclusion, "Gathering Connection," ties together the key themes of the book, reiterating the importance of the moral and material connections around water by concentrating on charitably endowed drinking fountains. Through everyday practices of establishing and maintaining *sabīls* to provide water to passersby, residents seek to build connections between neighbors (human and nonhuman) through acts of kindness, and at the same time to gather connections for themselves and departed loved ones that will benefit them in the afterlife. In this chapter I speak back to the state/development narratives of "traditional" understandings of water outlined in chapter 2, discussing the role played by water's divine provenance in how residents value, evaluate, and exchange it. *Sabīls* gather the central threads of the book, showing how everyday practices enable Ezbet Khairallah's residents to construct and manage connections materially, socially, and morally to their neighborhood and to each other in the ongoing project of being well connected.

The story of well-connectedness in Ezbet Khairallah is about how a resourceful but underresourced community manages a series of water-related conundrums. It is a story about how people make do around water and the particularities of the informal urban situation. It asks the question: what might we learn from the story of Ezbet Khairallah about how people live with water in cities? The elaboration of being well connected in this book lays out the constrained circumstances that shape people's actions. I also have attempted to capture a portion of the wit, wisdom, and labor that grow out of the overlaps of the frameworks for meaning-making and social connections around changing water systems. The institutional and organizational paradigms that work to shape the relations between water and people in Egypt are shifting. In tracing these changes ethnographically, I show the struggles over what matters about water. I also examine the stakes of different scales of analysis, which can hide or make visible information about how social and material systems intertwine.

Through the years, residents of Ezbet Khairallah have used a variety of mechanisms to improve the availability, quality, and timeliness of water services to their neighborhood and to clean their area of sewage. They struggle to remain hydrated, to rid their children of endemic intestinal worms, to avoid consuming water contaminated with sewage, and to mediate the impact of fluctuating water quality and chlorination levels in the larger potable water system in Cairo. These residents invest their labor to connect well through paying bills, forging networks of debt and reciprocity, and trying to make their land tenure claims stick. In a meaningful sense, the history of cities is about water, the future of cities is about water, and connections are the heart of city waterworks. Here, I tell one version of that story.

Chapter Two

Sowing Connection

During my first time riding the fifty-piaster pickup trucks from the el-Zahraa metro station to Ezba, I met Um Wasam and Um Kareem, who were taking their children to visit Fustat Park.[1] We became close friends,[2] and eight months later I returned with them to Fustat Park. On this return trip, our journey back to Ezba was markedly different from our trip out. Having spent the afternoon and evening in the park, we began our return around 8:30 p.m. by walking approximately half a mile from the park to the nearest intersection. We waited for about forty-five minutes there, looking for a bus or microbus to Ezba. We finally found a bus headed there, but it was jammed so full that there was no way we could fit in three adults and eight small children. Feeling responsible for returning them safely—this visit to the park had been my idea—I started flagging down taxis. The drivers all refused to go to Ezba. "Ezba? No, no."

As time wore on, I began offering an inappropriately high fare, but the taxi drivers refused. Finally, a driver in one of the new-style white, metered taxis stopped. He didn't know where Ezba was. I told him it was near el-Basateen and that we knew the road back. (I did not, but I assumed that Um Wasam or Um Kareem would know.) I told him I'd pay well, and we all began to pile in. I sat in front with one child and all of the bags, and the two mothers sat in the back with the remaining seven children. At this point, the driver started getting upset, asking me how many people were with me. When we finally got settled in, he said, "What, no one else? Fine, good. We aren't missing anyone."

It turned out, however, that Um Wasam and Um Kareem didn't know how to get back, having never traveled in a private car or taxi before. I remembered something about the way from our trip out via microbus and from previous trips into Ezba from el-Malek el-Saleh. I told him to take the left beside the big mosque, which he did, and then to go straight and take a right in a little bit.

DRIVER: No, left.

TESSA FARMER: No, left is el-Basateen, which I mentioned as being next to Ezba, but we want to go right.

DRIVER: No, left.

TF: No, right. See all of those lights on the right-hand side? Those are Ezba.

He gave me a considering look and continued on the road.

DRIVER: So, how long have you been here?

TF: Where, Egypt? About a year.

DRIVER: Where are you from? Lebanon? Sham [the Levant]?

TF: *Yaʿnī* [a nonresponse].

DRIVER: How do you like Egypt?

TF: I love the people. And these two ladies are two of the best mothers in the world.

DRIVER: And you've never met them before in your life, right?

TF: No, how can you say that?

DRIVER: You met them five minutes ago.

TF: Why would I go to Ezba with them if I'd just met them?

DRIVER: Don't you live there?

UM WASAM (FROM THE BACK SEAT): No, she lives in Hadayek el-Ma'adi.

DRIVER: So, you don't live in Ezba, but you know how to get there, and they don't?

We had reached our turn, so I told him to take a right after the cavalry station and follow the road to Ezba.

DRIVER: This road is closed, look at this road. Is it even a road?

TF: Yes, this is how we came down, we left from this road. It's open.

DRIVER: All of this for 15 LE? I'm going to have to pay 10 LE to wash the car after this. Isn't that right? Won't I have to pay that much?

TF: I have no idea how much it costs to wash a car.

DRIVER: What even brought you here?

TF: I have friends here.

DRIVER: Ah, but what brought you to visit this place in the first place?

TF: Ezba is nice and is getting nicer as they put in a sewage system.

DRIVER: But this place—

I cut him off because I could tell that insults to Ezba and to the people who live there were coming. I was embarrassed for him to say those sorts of things in front of Um Wasam and Um Kareem and wanted to defend them. But I ended up overshooting the mark.

TF: Ezba is a decent place, and the people here are *aḥsan nās* [the best people].

In response, all of the other adults in the car began laughing.

DRIVER: *Aḥsan nās?*

UM WASAM: And the best cars are the cars of Ezba!

UM KAREEM: And the best roads, the roads of Ezba!

TF: Why are you laughing? I don't understand.

UM KAREEM: *Maʿlīsh, maʿlīsh* [Never mind].

The absurdity of the moment was clear to them. As we drove past the guard at the gate to the cavalry station, his face showed his surprise at seeing one of the new-style white taxis drive into Ezba. Um Kareem kept pointing to the people who were laughing and pointing at us, enjoying the strangeness of the moment.

The difficulties of traveling in and out of Ezbet Khairallah, whether in microbuses, in taxis, or on foot, make clear the complicated position of this location, which both is and is not tied into the urban network of Cairo. This liminal condition is not coincidental: Ezbet Khairallah is an unwanted informal settlement that created the possibility of its own existence but not necessarily its permanence. In this chapter I examine how residents of the area sow the seeds of their neighborhood by laying claim to land, by occupying it, and by working to make it as comfortable and verdant as possible with their limited means and in a challenging context. I chose the term "sowing" because even in urban contexts, spaces require the right arrangement, timing, and balance of land and water to make life possible. Sowing indexes the work of keeping a thing alive in place, tending the possibilities for intergenerational flourishing (see Meneley 2020 on agro-activism). Land and water are connected in the acts of sowing and tending. In the story of Ezbet Khairallah, we see the deep ties between land tenure and utility access, and the link between both of these and the everyday work of watering relations.

I first trace sowing at the scale of the neighborhood as residents implant themselves materially, legally, and socially in their location. Sowing connection creates the possibility for being well connected. In this case, sowing "well" is about connecting deeply enough to firmly root the residents into this area, and reaching out tendrils to the adjoining neighborhoods to draw in the elements (like water) that are necessary for persistence in Ezba. "Connecting well" means resisting the forces that fight to keep this implantation unsettled, forces that work to uproot the neighborhood by creating sustained conditions of deliberate uncertainty. In this chapter I also trace early attempts to dislodge residents by demolishing their homes, as well as plans at the national and municipal levels to make financial hay from the valuable land on which the neighborhood sits and the residents' responses to these attempts to uproot them. I demonstrate the daily work residents do to sow their neighborhood with lively social connections and water resources. By connecting well, residents have grown a vibrant community on this contested and inhospitable land through everyday practices of obtaining, exchanging, storing, and utilizing water in the neighborhood—practices that involve extensive and skillful social labor and hard physical work. Despite broader narratives of the abjection of Ezbet Khairallah, I show that the settlement is tentatively on the rise, partly because of the beginnings of gentrification and partly because of residents' slow building of the possibilities of life.

Growing Roots

With an estimated population of 650,000 on two square kilometers of land, Ezbet Khairallah is one of the largest informal settlements in Cairo.[3] Greater Cairo

Ezbet Khairallah, 2010. Photo by author.

is home to roughly 17 million people, one of the most populous cities in Africa and the Middle East. Cairo contains approximately one-quarter of the total population of Egypt and half of the urban population (Sims 2012). During the twentieth century, much of the vertical and horizontal expansion of residential buildings in Greater Cairo was in unplanned areas, *al-manāṭiq al-ghayr mukhaṭṭata*, more commonly referred to as *ʿashwāʾiyyāt* (slums). In urban planning, these types of informal areas are distinguished by building patterns such as "small building footprints, 100 percent plot coverage and little or no allocation for public open space or social facilities" (Sims 2012: 62).

There remains a stigma attached to living in locations referred to as *ʿashwāʾī*, and some places, including Ezbet Khairallah, suffer the infamy of being marked as particularly dangerous. This was a running theme in my first months of fieldwork. When I talked to outsiders about traveling to Ezbet Khairallah, they often used a gesture of physical disfigurement or murder: a downward-slicing motion across the cheek or sideways across the neck to indicate the trajectory of a slashing knife. A ticket saleswoman in the nearest metro station used that gesture when she told me that the people of Ezba would kill me. The irony of the *ʿashwāʾī* stigma is that a sizable majority of Cairenes actually live in locations that could be classified as *ʿashwāʾī*. As of 2000, it was estimated that some 62 percent of Cairo's population lived in informal areas (Sims 2012: 69). Attitudes about Ezba and places like it speak to the fact that to grow roots in places marked as informal, residents must actively sow connections to the larger city in relationships that are both flexible and precarious.

In the twentieth century, there was a boom in informal settlement in Cairo. After the Second World War, people from the countryside were attracted by the growing industries in the city (Kipper and Fischer 2009). However, formal housing stock was in short supply. Rent controls and strict regulations on land subdivision and urban development, instituted largely in the 1940s, reduced availability. During this period, rural migrants largely settled in established neighborhoods in Cairo and worked to amass the capital that would eventually allow them to branch out into the city, buying plots of land elsewhere on which to build.[4] Informal housing at that time sprouted up primarily on agricultural land in the villages on the outskirts of Cairo (Sims 2012). Such land was attractive not only for its proximity to the urban center and lower prices, but also because historically these lands had been held primarily in small parcels of freehold property that were easily transformed into building plots. Lingering impacts of this history can be seen in the conversion of irrigation channels between fields into roads between buildings, as pathways for water in an earlier iteration of land use became pathways for people in informal settlements (Sims 2012; Kipper and Fischer 2009). In many informal neighborhoods in Cairo, people own their land, but they built the structures on it without official permission.

Later waves of migration, particularly during the 1970s and '80s, continued the expansion of housing on what had been agricultural land, despite countrywide legislation aiming to preserve farmland. This period also saw the development of many new areas on what was considered state land—primarily desert but also some state-claimed agricultural land (Singerman 2009).[5] Ezbet Khairallah occupies land officially designated as state-owned desert. Its streets and layout are therefore winding and unpredictable, unlike those of the informal areas built on freehold property in farming zones, which are based on grids structured by the irrigation paths that have been turned into roads (see Deboulet 2009 on the "dictatorship of the straight line"). As I saw when traveling to Ezbet Khairallah in the white taxi, the entire area—not just the entry but the larger geography—is marked as ambivalently connected to the larger urban infrastructure networks of Cairo.[6] Most residents of Ezbet Khairallah do not have building permissions for the structures they live in, and they also do not currently have legal land tenure.

Ezbet Khairallah, Cairo, Egypt, February 2021. Ezba is the area in the center surrounding the ring road and characterized by the nonlinear buildup of housing. Note the areas nearby with squared lines. Google Earth.

They are technically squatting on state-owned desert land, though there is ongoing legal action to gain land tenure rights.

Ezba, now a major informal area, has changed significantly from its earliest incarnation as a hilly tract of state-owned desert filled with unmarked graves and devoid of any utility infrastructure. Slowly, the neighborhood amassed: migrants from Upper Egypt and cities in the delta began building mud and rock homes in the area, and then people arrived from older areas of the city, such as Bulaq, el-Sayeda Zainab, and Bab El Shaaria. The land itself was parceled out by what both current residents and government officials refer to as "gangs," who carved out and sold plots to families seeking a foothold in the city. According to many of the older residents in the area, it was up to the individual families to ensure that their claim to that land "stuck," since they didn't have legal land deeds. This informality has had significant consequences on the development of Ezbet Khairallah: sowing connections has been a physically intensive project of squatting, claiming, negotiating, being chased off, and then rebuilding and reconnecting.

Hajja Mona, who sells vegetables in a makeshift kiosk, has lived in Ezba for roughly forty years. In the beginning, she said, elderly members of a family would stay at the house over the course of the week, while the adult children stayed closer to their places of work. On Fridays, the offspring would come and relieve the elders of their guard duty, and the elderly family members would go to their home in an older area of Cairo to take a shower and recuperate. (Those who came from Upper Egypt and did not have another house in the city made do with buckets of water for bathing, according to Hajja Mona.) Making housing claims stick meant not only having a permanent presence but physically marking the land as theirs through rock walls.

Agnès Deboulet (2009) discusses the relative positioning of the informal areas of Dar el-Salam and Establ Antar. According to Deboulet, Dar el-Salam, populated by lower-middle-class families, is "infralegal"—built without proper permits and despite laws prohibiting building on agricultural land, but purchased from landowning farmers and having long connections to state utility services. Establ Antar, which is considered by some residents to be part of Ezbet Khairallah, is less infralegal than extralegal or illegal. Establ Antar and Ezba more broadly are farther outside the lines than is Dar el-Salam, because Ezba residents lack both legal tenure and state services, and are primarily employed in the informal sector, Ezbet Khairallah and Establ Antar are thus marked as significantly more dangerous than Dar el-Salam.

The distinction between Dar el-Salam and Ezbet Khairallah is clear at the local level, but they are often lumped together under the designation "informal areas" when they are discussed in national and international circles. This, Deboulet (2009) argues, is a function of the metanarrative of informality and urban poverty that allows various agencies of the Egyptian state to "see without seeing,"[7] lumping these neighborhoods together as problems to be solved. According to this narrative, these problem areas are the product of social disorder, and in this way, informality remains the anomaly, the exception. This narrative elides informality's relationship to the inherent inequalities in the urban structure and ignores the fact that a majority of Cairenes live in technically informal areas.

Ezbet Khairallah residents' tenuous claims to their land are further threatened by urban land developers, who seek to uproot the people who have planted themselves there. The land on which the neighborhood is built was earmarked for luxury housing when President Anwar Sadat granted development rights to the Ma'adi Company for Development and Reconstruction (MCDR).[8] The area, which rises along a ridge of limestone near the Nile River, is only five kilometers south of downtown, making it an ideal location that provides quick access to many important historical sites. However, MCDR's development plan faced pressure from the increasing numbers of informal claims to the area.

MCDR also faced challenges from the Ministry of Antiquities, which in Egypt is a major political force: antiquities are a large draw for tourism in the country, and tourism is a major sector in the Egyptian economy.[9] Cairo is an old and storied city, having first been settled by the Fatimids in 969 CE. What is now considered Greater Cairo absorbed the earlier capital cities of Memphis, Giza, and Fustat (Abu-Lughod 1971). Fustat is very close to Ezba, and that proximity has left lasting impressions on both locales. Parts of the area that is now Ezbet Khairallah remain under the ambit of the Ministry of Antiquities, which is charged with preserving archaeological sites, managing research on these locations, and developing them for tourism. Like MCDR, the Ministry of Antiquities seeks to uproot residents of Ezbet Khairallah and sow its own connections in this piece of land.

A contact in the Ministry of Antiquities described ongoing attempts by the ministry to protect relatively small parcels of land around archaeological sites in and near Ezbet Khairallah. Within the borders of Ezbet Khairallah are several lesser known archaeological sites, including Sabaʿ Banat (Domes of the Seven Girls), the Ahmad Ibn Tulun Aqueduct, and the Muhammad Ali Gabalkhana in Establ Antar (Tadamun n.d.; see also Dhahir 2010). To the north are the remnants of the city of Fustat, and to the east are the southern cemeteries. Many residents spoke with me about finding skeletal remains during the construction of their homes. They had varying ideas about the era of origin for these bones. Many of the skulls had tendrils of red hair, suggesting to some residents that the bodies were English soldiers from the time of the Second World War, but hair reddens during the process of decomposition, and the color likely had nothing to do with the heritage of these skeletons. Others pointed to the nearby tombs in the southern cemeteries and suggested that the bones were the remains of people too poor to have merited mortuary structures.

Whatever the provenance of the remains, several Ezba residents reported encountering bodies while constructing homemade septic tanks and water pipes, and they carefully reburied them somewhere safe. One resident spoke about her sense that in trying to build a life in Ezbet Khairallah, she was growing among the dead. Much has been written about residential neighborhoods constructed within Cairo's City of the Dead, but in Ezbet Khairallah, the reburied dead nourish the living earth, fertilizing the ground to grow a new neighborhood.

Land and Water, Tenure and Utilities

Plants begin with sowing, but they are tended through watering at the right time and in the right quantity. Watering in Ezba—creating infrastructure to obtain and transport potable water—has been a difficult, long-term struggle for the residents. When the area was first settled, water had to be bought from mobile water carts or trucks or carried from public taps in neighboring areas. For the first couple of decades, women in Ezbet Khairallah walked to and from the public taps in Abu Ashraf and el-Basateen, balancing jerricans on their heads. Several longtime residents told me that the lines at the public taps were very long, and it could take hours to get one's turn. Women would sometimes send a child with an empty jerrican to wait in line, joining them later to carry back the full, heavy container. Sometimes, people would leave buckets by themselves to keep their place in line, but that invited line jumpers, and one woman told me that she had seen several physical fights between women in the lines. As a young teenager, she had been afraid to fetch water by herself; it was better, she said, when you could collect a few neighbors together to go at the same time.[10] Many of the women who lived in Ezba at the time said that the people of Abu Ashraf and el-Basateen did not like the people from Ezba coming in and crowding the lines at their public taps. These conflicts over water exacerbated tensions between the areas over the perceived lawlessness of Ezba residents. Eventually, most Ezba residents built their own utility infrastructure, installing unauthorized potable water pipes and making illegal electrical connections to adjoining neighborhoods.

Sara is a young woman who came of age after my earlier period of research (2009–2011) and with whom I reconnected on my return in 2018. I asked Sara about the changes that she had experienced in water systems in Ezbet Khairallah over her lifetime. Although her reflections about which stages were the most difficult differed from those of older generations, who had longer memories of life before there were unauthorized water connections, she had similar stories about the deftness of social labor and the difficulty of physical labor in ensuring a sufficient household water supply.

Sara told me, "There were struggles. You know what I experienced? I didn't experience filling water from Basateen. I didn't experience that, to be honest. But I did experience when the water was cut very often. More than you would believe." The cuts in service were frequently due to the maintenance of pipes or repairing damage, sometimes from construction in nearby areas. Getting the water that Ezba needed in order to grow was a long process that was neither linear nor consistent over the area. Houses gained and then lost access to water; some houses and some sections of the area delayed their installation of unauthorized water systems; and increased access for some areas sometimes reduced the supply available in other areas due to inadequate water pressure. It was not always clear, however, why there was no water. Sara explained:

> But, for us here at first, we would come in the morning, for example, we'd get up for the day, open the tap, and not find water. It might stay that way until the next day, with no water at all. What would we do? There is one here—do you know Um Hameda, the lady with the *kūlmān* that you spoke with—she would often have water. And we would fill water [containers] from her [house]. I would fill a jerrican, or fill a small washing pan, carry it, and come [home]. One time I would carry, another time Momma would carry, and we would walk home. Momma would carry a big washing pan. At other times—do you know the mosque just over here? [gesturing] Behind it is a parking place. In that place there was also a tap that was open to people. People would come to fill up.

Sara's description of navigating uncertain water access illustrates how integral her social relations are to her capacity to procure water. Not only are her intimacies and positive connections with multiple people necessary, but those relationships must be located in the right place. Sara connected well in the public parking lot behind the mosque and in Um Hameda's house. She also connected well in taking turns with her mother to carry the water home. Connections are sown using social labor to keep sharing economies lively and with physical labor to carry burdens over distances. Being well connected in this context does not mean that Sara has strong relations to the state nor that her water supply is flawless. Instead, it means that she has and continues to sow spatially and temporally specific ties that enable water to reach her house via a complex array of possible means.

Residents didn't always know why, when, or where a cut in supply would happen, and houses were sometimes givers and sometimes receivers of shared water. Women in the neighborhood developed a general knowledge of the main junctions in the unauthorized piping system near their homes. When a cut occurred, they visited neighbors and acquaintances in their social networks to see who still had access to potable water or who knew where water was still flowing. Sara's mother made clear the elliptical nature of the water supply in Ezba. She told me that the house where she was born, near the edge of Ezba, had been one of the first in the area to have a formal water connection, and her household had sold water to their neighbors. Some people in Ezba had come and purchased water occasionally, while others paid a monthly fee for regularly filling up from their home. Some water was transported by being carried in small pans, while nearer neighbors had hoses, which they connected to her faucet and then filled storage containers in their houses. For Sara's mother, connecting meant having a sound water connection, which gave her the opportunity to be a taproot for others, who carried water from her house to their own. Part of sowing connections in Ezba is participating in a particular kind of sharing that is oriented toward ameliorating the uncertain and uneven ruptures in the physical architecture of water connectivity. Here, intermittency defines the stakes of connecting well.

The times that Sara remembered, after unauthorized piped networks were largely in place, and the routinized crises of access described by Sara's mother wore different faces. In Sara's experience, people drew on their own reserves of stored water or attempted to locate willing neighbors whose pipes were still running. Because demand was intermittent, this type of water sharing was often free of financial costs, but it had steep physical and social costs. Sara told me, "In those days, I would go to carry something on my head. The water would splash down over me. [laughter] I told Momma that I was soaked through. Because I was carrying it on my head." In retrospect, this story was funny, but the physical demands of carrying water that it encodes were very real. Similarly, these transactions required smart social labor, for residents had to negotiate the daily details of collecting water, negotiations that could include conflict. Sara continued:

> In those days people would fight with me. Sometimes a girl would say, "I'm going to go, I'm going to go [to cut in line]." This sort of story happened a lot. . . . A girl, for example—I used to be—not in the past, I am still—easily irritated. [laughter] I hated for anyone to make me mad. I was holding [a container] and entered with it. I had been standing there for a long time. She said to me, "No, I'm going to enter!" She looked scary, like maybe she could hit me like it was nothing. I told her, "No, I'm going to enter. I've been here for a long time." She said, "No, you're not going to enter!" and she really yelled at me. I tried to push her [away]. I tried. I found that she had grabbed the washing pan I had, it was a small one, and she threw it at me. She pushed me and told me, "You aren't going to fill up at all."

Sara and I laughed again, and she continued, "Then I went and got my mom for her. I told Momma, 'Come see, this girl hit me.' My mom yelled at her. The girl's mom yelled. It was a big fight. And in the end we came [home]. It was a really difficult issue. We had to exhaust ourselves to get water. And when the water came, we filled everything. We filled everything in the house. After we filled everything in the house a little while later, the water would cut again. Every little while. My goodness!"

Sowing connections not only is about creating harmonious or smooth social interfaces between people, but also includes conflict in order to maintain connections sown to places, including access points for water. A romanticized perspective on sharing economies or collaborative survival practices does little service to understanding how the "well" part of connecting well is made possible (see Wutich et al. 2018). As this story shows, exhaustion is a key part of social labor, which is never without accompanying physical expenditures. Navigating austere conditions of disconnection requires the managing, mitigation, and creation of conflict. In the face of conflict, smart social labor often means reaching tendrils out to other networks—in this case, each girl's mother as members of family networks—in order to expand audiences of accountability.

In this story, the exhaustion, both physical and emotional, was palpable to me as Sara spoke, because her story did not register the fight as an event with an end. Instead, her story rolled into a continuous pattern: "After we filled everything in the house a little while later, the water would cut again." What is sown into the fabric of her tale is not only her tactic for insisting on rootedness to place, in this case via a fight, but also an integration of exhausting social labor into the narration of time's passage, which is measurable in patterns of disconnection (water being stopped from flowing, a girl stopping her from entering) and attempts at reconnection.

As these examples demonstrate, getting water in Ezbet Khairallah at the right time and in the right ways fell differently on each household at different times. Some had official water, some had unofficial water, some had disrupted water, some had no water at all. The labor of managing water also fell differently on different members within the household.[11] Globally, the task of fetching water disproportionately falls to women and girls. There are physical and social costs and also high economic costs.[12] My interviews and participant observation in Ezbet Khairallah largely (though not always) bore out this gendered pattern. Sara said that her younger brother would come to help carry water: "He was four years old. There was this little, tiny jerrican. I would tell him, 'My love, you won't be able to carry that.' He said, 'No, I want to carry it.' He loved to help, just like Momma [who was well known for helping others with physical labor]. And he would walk back with the jerrican." In Sara's family, then, fetching water helped to cement their ties and connections, as members of the family helped each other.

The intermittent supply also brought to the fore questions about the quality of water, which Sara highlighted: "When the water would cut for some hours or maybe a whole day, when you went to open the faucet to fill something, you would find that the water would come out black. . . . Why? You left the pipes empty for a while. I don't really know what happened to the pipes. Momma always used to say, when the water cuts and comes back, open the faucets and leave them. It would wash the pipes clean. I'm not really sure [how]." Sara then reflected on how the present moment differed from the past that she was recalling: "Nowadays, the situation has changed, to be honest. It's not like it used to be at all." "Thankfully," I responded. She continued, "Now, if the water cuts, it doesn't take a long time [to come back]. Or, if it cuts for a while, the water doesn't come out dirty in the same way. Not like before." In 2018, when this conversation took place, the events in which sewage water infiltrated potable water pipes in the area had significantly declined due to the completion of the municipal wastewater system, but the problem had not completely disappeared.

Land tenure and the history of informal settlement in Ezba have shaped the way access to water and related utilities, like electricity, is managed. Cuts to supply reflect how physical connections to formal infrastructure differ meaningfully in places marked as nonplaces. The historical dilemmas of water collection and transport described by Sara have been caused by Ezba's relation to Greater Cairo and by the increasing complexity of its residents' land tenure arrangements.

Ustez Ali has lived in Ezba for about twenty-five years. During the first eight years, he said, there was no electricity, and water was difficult to come by. His house now has official connections to both systems, but when we spoke the bureaucracy around utility billing was causing him grief. He told me about an electricity bill that they had received for the previous two months that totaled more than EGP 300. His wife, Um Farha, had gone to the nearby electric company office to argue that they couldn't possibly have used that much electricity in a two-month span. The company's representative told her that the billing rate and system were set by the government, and there was nothing she could do about it. Um Farha would just have to pay. Um Farha stressed to me that they didn't have much in their home that ran on electricity, just a few lights and a TV they used a few hours a day. A shop was built into the first floor of the family building, but it had long been effectively closed. Ali speculated that someone from the government just looked at the registry, saw that there was a shop, and assumed that it must use a lot of electricity. "Rich people can do what they want here, but poor people can't do anything or even say anything about water or bills or the system. They will send in the national security police. Everybody in Ezba is just saying 'yes' to the government, whatever they tell us."

Although Ali felt that the government just estimated how much should be paid (an estimate that may have had nothing to do with the reality of his family's life), there was no real choice. If they didn't pay, the utility company would take the meter, leaving them without electricity altogether. This was the flip side of having formalized the household's electricity connection. Ali was able to get the electric meter installed because a political candidate had said that those who signed up to vote for him could do so. Ali signed up to vote on the strength of that promise; he wanted a formal electricity connection as part of a strategy to get the state to recognize his land tenure. Ali told me that he paid 5 percent of the value of his land to the government for his occupancy of it, but the government still classified him as a tenant of a "condemned" building or ruins, not as an official resident landholder.[13] He said that his family still had no real rights to the land.

As Ali's relationship with a utility infrastructure illustrates, formal connections sown to water or electricity are often double-edged. They render people and homes in Ezba visible, but at a high cost; sometimes there is a net gain over an informal connection, and sometimes there is not. As seen in Ali's story, the provision of utility services in Ezba has sometimes involved votes-for-services swaps. First, community members petitioned the state through demonstrations and public gatherings to get services. In response, political candidates promised to implement first electricity and then water utility services in exchange for votes in parliamentary elections (Blaydes 2011; Chatterjee 2004). For years, this dynamic even played out in the pipes: there were preelection surges in water pressure (Albuquerque 2010). Before elections, strong flows during daylight would replace weak trickles at odd hours, changing the pace and smell of life.

As people try to stay in place, they negotiate disruptions to the streams of resources that must flow in order to keep life going—water, electricity, and sociality. When connections are too intense or too costly, or when they are cut or diminished, life is harder. In the context of informal and partial land tenure and utility access, sowing connection is about getting the right combination of water and land, at the right time, to make life possible. People in Ezba, especially women, navigate contradictory historical developments, trying to stay in place by sowing into being that which is constantly undermined, sometimes by paying for the privilege of using the land they live on or by agreeing to accept the cost of utilities moving forward—all in the hope that these negotiations will lead to stability. Their hope is not a blind one: smart social labor includes making good connections with entities far outside the boundaries of Ezba.

In 2010 when I asked people what they thought Ezba would be like in a few years' time, many said that they thought it would be the best place to live in Cairo. It is only five miles from Tahrir Square and near to the metro and to Salah Salem Street. The future looked bright—if utility services continued to improve, and if the people who lived there would still be around to enjoy the benefits. Residents of Ezba sought to ensure these outcomes. They created advocacy communities centered on issues of wastewater management, partly in response to the 2008 el-Duwayqa rockslide, which had been caused partially by sewage overflow in a similar informal "slum" area. But residents of Ezba largely centered on using collective action—their connections with one another and their social labor—to change their own situation. People marched. People protested. People sat on the doorsteps of ministries and didn't go away. As one longtime employee of the Ministry of Water told me, "Those Ezba people came and kept coming, and they would block you from getting into the ministry. Eventually, the ministry just gave in." Perhaps "just gave in" is not entirely accurate, but the government was certainly motivated at least in part by public pressure. This pressure was intensified by residents' savvy leveraging of new media. When I asked one Ezba matron how the neighborhood got approved for a state sewage system, she said that they marched and embarrassed the government. "You can see me on Dream TV at the marches," she told me. "Look it up on the internet—they have everything there."

Risky Locations: Challenging Implantation

Sowing connection in Ezbet Khairallah has been a painstaking and challenging endeavor for residents, primarily because others do not want them living there. As mentioned, the Ministry of Antiquities and the MCDR both wished to sow alternative projects on that land, and residents of adjoining neighborhoods (also largely lower-income areas) were not happy to have a "dangerous" place, known for crime, nearby. These people and organizations often worked in coalition with the Governorate of Cairo (GOC) to uproot what Ezba residents were working to grow: on multiple occasions, the GOC sent in security forces to evict the inhabitants by demolishing homes.[14] Sara's mother, Um Sara, reported that her family's natal house was destroyed three times—and rebuilt three times—over the 1990s. Residents worked together to rebuild homes each time and to pursue litigation for residence and land tenure rights.

In 1983, following a clearing attempt by the GOC in 1982, a local textile merchant named Sikh Ahmed Muhammed Gahanna led a class action lawsuit against the governorate (Tadamun n.d.). In 1984, they won an expedited court ruling that stayed a demolition order, and residents organized to build more utility connections and negotiate with utility companies to pay the fines for existing unofficial infrastructure. In 1986, a new legal suit for land tenure rights was submitted to the administrative court, but in 1987 the lawsuit's request for residents to be able to buy their land officially was rejected. That ruling was overturned in 1999 by the supreme administrative court, which ended the MCDR's rights to the land, made the GOC responsible for providing utility services, and said that the GOC must sell residents their land. However, the ruling did not specify a price, and the GOC said that it had to have a plan for the area before a price could be set for land purchases.

In March 2011, a new organization, the Khayrallah Lawyers League for the Defense of Rights and Freedoms (KLLDRF), was founded, and in 2012 it brought a lawsuit on behalf of one resident. Ultimately, that case was found in favor of the resident's right to purchase the land from the GOC, but the ruling stipulated that the land had to be purchased at *current* market value. The KLLDRF filed an appeal demanding that the value of the land be priced based on the time of adverse possession (or squatting). When ruled upon, this claim, which references a 2006 law on land tenure in Egypt, will set a precedent for price. The issue is still pending today, and residents continue to fear that they will be removed from the area because they have not been able to purchase their land officially from the GOC. This fear has waxed and waned over time but became more intense during my research period because some houses along the cliff were deemed unsafe and removed and also because of the el-Sisi regime's accelerated attempts to reclaim state land throughout Egypt. As Ezba has gained more and more residents, the area must be reckoned with.[15]

The kinds of risk and uncertainty associated with Ezba crystallized in the fallout over the el-Duwayqa disaster of September 2008, in which 119 people were killed and many more wounded by the collapse of 18,000 tons of rock from a cliffside onto an informal area below in Mansheit Nasser (Amnesty International 2009). In the immediate aftermath, roughly a thousand buildings in the area were destroyed, and people were forcibly relocated. The story of the disaster focused on blaming overflowing septic tanks from informal housing on the cliff, which had caused layers of clay to expand, putting pressure on the limestone layers that comprise the foundation of the hill (Amnesty International 2009: 16). Of course, the area was already at risk from the natural geography; residents had been warning the Ministry of Housing and local elected officials about the danger posed by rocks that hung perilously over their homes, but they had received no reply. Septic tanks, however, were only one piece of the puzzle—albeit the piece that received by far the most attention in the national and international press. Another rarely mentioned factor was the construction of luxury housing atop the Mokattam plateau: the Uptown Cairo project by Emaar Egypt, a subsidiary of Emaar Properties, based in the United Arab Emirates. The use of extensive amounts of water to prepare the land for building and for water features in landscaping and golf courses added additional pressure to the already-eroded limestone foundation (Hussein 2008).

When the rockslide took place, the poor neighborhood's informal status affected the official response. Rescue workers' slow and incoherent response caused problems, which were compounded by the police and military's refusal to let in voluntary assistance from other sectors. Many people who lost their homes were moved to Sixth of October (one of many desert satellite cities that have been built radiating out from Cairo). Some were forced to sign contracts for resettlement apartments that they did not receive. Others, including some large families, were housed in small rooms without services in mining camps. Hard on the heels of the rockslide, houses near el-Duwayqa were demolished by the government—largely with their contents intact, as people were told to evacuate immediately and had no time to save their belongings, either during the emergency or afterward (Afify 2014).

A month after the rockslide, Mubarak issued a presidential decree establishing the Informal Settlements Development Fund (ISDF), which was tasked with improving slum areas on a national scale. The first step of the ISDF was to survey all of the *ʿashwāʾiyyāt* and categorize them as either "unsafe" or "unplanned." Within the unsafe category, there were three degrees of seriousness. The first degree meant that an area posed immediate danger to the occupants, who had to be moved straightaway.[16] The second and third degrees of seriousness involved unhealthy situations or areas facing some kind of specific threat, such as industrial pollution or exposure to high voltage power lines. Some unsafe areas (mostly those in the first and second categories) were slated for immediate demolition. The category of unplanned usually indicated that occupants lacked building permits and/or tenancy.

In Ezbet Khairallah there were two locations that fell into the first degree of seriousness: the area along el-Negh Street overlooking the cliff and portions of Establ Antar. With little notification, buildings along both of the cliffsides were destroyed, and people were forcibly relocated to Sixth of October City, far removed from their social networks and places of work. Some subsequently returned to the area, but they had been transformed from modest homeowners to anxious renters or couch surfers, staying part time with friends or relatives in the area during the week so they could work (Al-Sharqawi 2010).

The ISDF was established with the guidance of Dr. Ali El Faramwy, who was the Egyptian national program manager at UN Habitat at the time of the el-Duwayqa rockslide. When I spoke to El Faramwy in September 2010, he said the ISDF was basically an implementation of UN Habitat guidelines with minor tweaks. It functioned essentially as a bank to lend money to local governorates, which were then supposed to use the funds to build new housing on half the land and sell the other half to repay the building costs. This formal and official relationship between international capital and state governance structures has intensified the structural inequalities facing urban poor people in Cairo. The relocation of Establ Antar and el-Duwayqa residents to the desert satellite city of Sixth of October has reverberated as a dire warning of a possible future for Ezba's residents if the land that they sit on becomes too valuable.

Valuable Land

The ISDF scheme for managing unplanned areas was explained to me by Dr. El Faramwy. I met with him in 2010 in the fund's office, which was located in an unmarked building with a guard at the gate. The building was pleasant but not sumptuous. El Faramwy was very familiar with the informal areas that are the subject of the ISDF's work. He knew the geography of Ezba and could describe the sites where people had been moved. The ISDF built on the Development of Informal Areas Project (1994–2004) under the Ministry of Local Development. That project had provided infrastructure like water, roads, sanitation, and electricity to informal areas. In 2004 the project moved from the Ministry of Local Development to the Ministry of Housing and Urban Development. The Ministry of Housing chose to focus primarily on landownership. The project's improvements to unplanned (not necessarily unsafe) areas were not the central focus. These informal neighborhoods needed legal documents and detailed plans to support landownership cases, but the legal documents couldn't be issued to *ʿashwāʾiyyāt*. So *ʿashwāʾiyyāt* were left alone.[17] That was the situation from 2004 until September 2008, when the el-Duwayqa rockslide happened, and a shift in focus was made from infrastructure to building houses.

The fund is designed so that projects begin with a request from the governor of a district, who presents the ISDF with a detailed action plan that meets specific requirements; these requirements are modeled on UN Habitat documents on slum upgrading. Local actors must have all the legal documentation complete and a fully acceptable action plan in hand. The fund then lends the governorate the money, which the governorate uses to upgrade the area and then pays back over time. According to El Faramwy, part of the innovation of the ISDF in Egypt is that the plan is based on local, decentralized planning.[18] The whole project is set up on a cost recovery basis or, as El Faramwy put it, land-based finance.[19] The idea is that the land belongs to the government anyway, but it can't access the land because it can't move people who have built housing on it without giving them somewhere else to live. So, the fund proposed that the government take control of land on which people have built one- or two-story homes, clear it, construct a five- or six-story building of apartments on half, and then sell the other half through a commercial auction run by the GOC. Then the GOC would use the money from the sale of the valuable land to pay back the loan used to fund the construction of the apartments. Any excess profit from the sale of this land should then go to cover the expenses of clearing areas of the first degree of seriousness and for social programs named in the project plan.

TESSA FARMER: Is it literally the same land? So, if one *feddan* [0.42 hectares] was recovered, half would be sold?

EL FARAMWY: Yes, say you recovered ten *feddan*. You would build apartment buildings on five *feddan* and sell the rest. In the package for the development of the area is the whole network of water, sanitation, electricity, etc. Also included are monies for social development. The fund provides money for women's health, youth employment, adult literacy, and national ID cards. This is paid for by the land sale.

TF: Do the houses belong to the people of the area?

EL FARAMWY: No, they are rented apartments.

TF: Will the same people be moved back in who were living there before?

EL FARAMWY: That's all up to the local governments. The process is local and decentralized. Local governments get together with their stakeholders, NGOs, local representatives, and decide what's important to them. The General Organization for Physical Planning prepares strategic plans in cooperation with the local governments. The way it used to work was that the Priority for Cairo [project] would pay 15,000 to 25,000 LE of the cost of building an apartment that maybe cost 75,000 LE. The local government would pay the rest. Then, if the tenants didn't pay their rents, it became a drain on the revenue of the local government.

This plan articulates the source of Ezbet Khairallah residents' anxieties. If these plans came to pass, the homes that they had built and rebuilt through laborious and long-term investments of financial and social capital, the neighborhood connections that they had crafted through careful social labor, and the roots and stems of their moral engagements with place would be severed (Mohamed 2010). Residents would become unwanted growth to be plucked from the soil and replaced with better seeds. Nor was this the only plan for massive revisions of the area on which Ezbet Khairallah rests. The Cairo 2050 master plan for urban redevelopment in Egypt's capital city shows the area of Ezbet Khairallah as completely removed, though that plan saw uneven and slow implementation both before and after the 2011 revolution.[20] This tension was evident in 2009–2011 as residents watched the implementation of the neighborhood's sewage system. They saw the connection to formal state water systems as a potential first step in their removal from their homes.

From this perspective, when the liberal paradigm of state provision of water services is extended to a previously unserved location, it gains a new valence. It looks less like benevolence than like danger, with water not as a human right but as a wave that might wash people away, clearing the land for greater capitalist accumulation. This is the mode articulated by Harvey (2003), who traces the ways that city spaces are particularly valuable or available to accumulation by dispossession; capitalist market structures allow wealth to be built through the destruction or capture of other people's resources, including real estate. This is part of the anxiety of Ezbet Khairallah as a location. Residents recognize that their presence is unwanted, and they see the implementation of the utility infrastructure as creating the possibility of their dispossession because utilities make the area a better place to be. The more the state knows about them—the number of inhabitants, where they live, who claims what land, the dimensions of their houses—the more it can use this information against them, including making it available for projects aimed at uprooting them. As Antina von Schnitzler (2008) traces in the context of South Africa, water meters enable mechanisms of accounting for and particularizing populations, articulating a new, rational version of citizenship; thus, the implementation of water infrastructure has the potential to damage humans and their well-being. Ezba residents are anxious because they see that the state's extension of water and wastewater services to their neighborhood is a potential mechanism for enacting their displacement.

I heard this anxiety firsthand in my conversations with Um Ameer during the sewage system installation in Ezba. Indeed, a form of displacement began during the installation itself: at least half of the length of her street was torn up, and workers had not left much room to walk along the side. Wooden planks were placed there, but the improvised walkways were narrower than usual, and a backhoe was perched at the near end of the excavation in the middle of the street, blocking most of the path. Two men moved out of the way to let me go by, but one muttered after me, "If you are even able." I made it down the street, then turned around and came back up along the other side to reach Um Ameer's apartment building. Her aunt had temporarily closed her small street-level store because of all the work in front of their place. When I arrived in the apartment, I was out of breath from climbing around the piles of dirt. Um Ameer went onto the balcony to take a look outside at all of the goings-on. I followed her, and we both watched the activity below.

I asked how she was managing with the street closures. She told me that because of the hubbub, she hadn't been out of the house since the day two weeks before that I had last been to visit, when the kids were going off to their grandmother's. She noted that the workers were having a particularly bad day today and that the mass of thick, dark water filling the trench was from a septic tank that they had punctured while digging the ditch. The ditch was almost full, and they had put in a pipe and pump to drain the dirty water into the main street of Ezba—a river of sewage flowing on the central thoroughfare. While we stood there, Um Ameer told me that she had heard that the government was going to move people from Ezba to Sixth of October City, as they had done with Establ Antar residents whose houses were destroyed. Um Ameer said, "Maybe it won't happen. Maybe they will only take the people near Ain el-Sira that they are afraid will fall into the water. Like they did with the people whose houses collapsed. Did you hear about that? They may take them for their own protection."

I replied, "But isn't it weird that they would put in all of the services and then move people? Especially as they are putting the services in along the routes of the streets as they are now [meandering]?" Um Ameer repeated, "Maybe it won't happen."

Her sister-in-law, Um Noura, arrived with her little girl and joined us on the balcony. While we were standing there, there was a commotion among the workers. They had now struck a water main, and fresh water was running in the ditch along with the dirty water. The workers were worried that the ditch would overflow and flood buildings along the street. When they finally got ahold of the man in charge of the water main, however, he refused to turn off the water. Um Noura explained to me, "Everyone in Ezba drinks, cleans, does dishes from this water. People need to pray [the] Maghreb [prayer]. What can he do? He can't turn off the water." Um Ameer responded, "All of Ezba will be full [of water] if they don't close it."

Below us, we could see a group of men, one looking very official in a suit and holding a clipboard, arguing with a group of workers in "professional" clothing while another group of workers in overalls stood around and watched. Finally, some sort of decision was reached, and a worker was sent to adjust the pump that was draining the sludge. He used his bare hands while standing on a very thin wooden plank over a pit of sewage water. More water drained into the main street below.

Um Ameer's sister-in-law turned to her and said that it didn't matter. They were all going to be moved to Sixth of October City, and Ezba in its entirety was being sold to some rich guy.

> UM AMEER: They are only going to move the people by Ain el-Sira, where they are afraid that they will fall into the water.

SISTER-IN-LAW: No, they're going to start up there, by the church, and then take all of Ezba below. We are all going to be moved.

UM AMEER: Then why would they be installing all of this? Why all of this work on a sewage system if we are just going to be moved?

SISTER-IN-LAW: Those aren't sewage pipes! Haven't you seen those pipes? Look below you! See those pipes? They are gas pipes. Haven't you seen the trucks coming and going, coming and going from Ezba? They aren't installing sewage systems for us! They'll leave Ezba for five years, then take us all out to the desert, to the far Sixth of October.

She looked to me for sympathy—to see if I recognized the disaster that this would pose for them. The removal of their community from Ezbet Khairallah would not only mean the loss of the wealth tied up in their ownership of their land or apartment, as laid out by El Faramwy of the ISDF. It would also mean the loss of their social networks, their connections to jobs and resources in neighboring areas of central Cairo—networks that David Sims (2014) and Julia Elyachar (2005) argue are one of the key mechanisms through which those on the urban margins can survive. Infrastructural connectivity is potentially deeply disruptive and disorienting to these crucial social connections. Ezba residents had already witnessed the plight of those left homeless from el-Duwayqa, and they lived every day with the implications of being in an area that the government "sees without seeing." It was hard for residents to trust the intentions of the governorate. The backhoes and heavy equipment rolling in to Ezba conjured dystopic images of destruction and displacement for people who usually found themselves at the receiving end of state security forces rather than state welfare services (Peteet 2017).

Propagating Community

Everyday practices are the building blocks of being well connected in Ezba, as residents strive to create a network of neighborly relations and organizational systems that enable them to manage shortages and get daily care tasks accomplished. As Sara and her mother told me, in the early days, women queued for hours to fill jugs at water taps in adjacent neighborhoods, built friendships, engaged in fights, carried water on their heads, shared or purchased water from neighbors, and bought water from passing vendors. Over time, residents of the area connected to the formal potable water networks in neighboring areas, sharing the cost of illegally building offshoot potable water mains. Some water connections and additional water pressure also came through votes-for-services swaps—a common transaction in Egyptian electoral politics and elsewhere (Anand 2017; Piper 2014; Sims 2012; Chatterjee 2004).

As informal connections to neighboring water systems became prevalent, residents worked to address the gaps in their access: times when water was cut, incidents when there was insufficient pressure to reach homes atop the plateau or apartments higher up in buildings, limits to supply created by challenges in managing grey and black water, and leakage of sewage into potable water systems. Residents called and visited water complaint centers, woke early or curtailed travel in order to fill bottles and barrels of water during times of access, shared with neighbors, and installed water pumps in apartment buildings—all to make it possible to clean their homes, cook food, and organize their households.

Water management made up a large part of the women's domestic labor, labor that has long been a topic of scholarly interest in Cairo (Deeb and Winegar 2012). In the 1990s, research in Cairo dealt largely with questions of domestic and small-scale economies in popular urban neighborhoods, especially as these issues related to gender.[21] Homa Hoodfar and Diane Singerman (Singerman and Amar 2006; Hoodfar 1997; Singerman 1996; Singerman and Hoodfar 1996) have examined in great depth gendered norms of labor, marriage choices, survival strategies crafted around microlending, food distribution, and the management of intrafamilial strife in poor urban areas of Cairo. Hoodfar (1997: 55) sees her work as filling an empirical and theoretical gap in scholarship on household economies in the Middle East, studying men's and women's decisions about marriage partners; the allocation of labor between domestic work and employment in formal and informal sectors of the economy; consumption practices; and fertility preferences. In her view, these are all carefully considered strategic choices that maximize limited resources for low-income urban households.

Singerman also examines the political and survival methods of urban poor people in Cairo. In her book *Avenues of Participation: Family, Politics, and Networks in Urban Quarters of Cairo* (1996), Singerman attempts to address a theoretical and methodological oversight in the study of politics in the Middle East, which has focused on formal political associations and the political maneuvering of elites. Singerman instead studies the politics of everyday Egyptians as she works to reformulate the traditional understandings of what the "political" is. She argues that the activities that typically define the political, in which rights-bearing citizens organize in open forums to lobby the government for their rights, is inadequate to deal with the complexity of everyday political activities (see also Adely 2009).

This is, Singerman says, particularly true in Egypt for two possibly linked reasons: political repression in that nation is so severe that only a powerful few can openly oppose the state, and poor people must therefore use other avenues to seek their interests; and the household in Egypt is much more integrated into the public field and national economy than has been recognized in typical public-private distinctions in scholarship on politics in the Middle East.[22]

As these previous studies of urban domestic life in Cairo show, the domestic is not a stable category; it is produced in conjunction with other realms of action, from the public to the political, in part by women's smart social labor and hard physical work. As Elyachar (2010) points out, in Cairo (as in many locales), women's labor is an often-unexamined foundation for constructing properly social family members. This female social labor ranges from sending children to school with brushed hair and men to work with ironed clothing to forging and maintaining the social connections that produce potable water in the household.

Maintaining household cleanliness was a particular challenge in the saturated conditions of Ezbet Khairallah when septic tanks dominated waste management, as one interlocutor pointedly demonstrated in a conversation early in my research. Um Mina drew her chin back, pushed her cheeks out, and looked at me from the corners of her eyes: "I clean it every day. I'm very clean, very clean. I put a lot of water on the floor, a lot! I cook every day. I don't serve leftovers. And I clean every piece of food that comes into my house." She invited me into her kitchen, and I watched her clean meat, pouring dish soap into a plastic bowl in the sink, kneading the chunks of meat like dough, and squishing it between her fingers. At the same time, her daughters pulled all of the rugs off the floor to demonstrate the vast quantity of water that they used to mop and clean the apartment. Buckets of water from the bathroom tap were hauled into the room and spread around with squeegees on a broomstick. Long strokes moved water into the corners of the room, and rapid pulls of the squeegee moved the water to the bathroom, where it disappeared down a grated hole. The girls had rolled up the hems of their athletic pants, a common outfit, to work the water thoroughly over the tile floor.

Um Mina's refrain—"I'm very clean, very clean"—opened and closed every conversation that I had with her. For all the months that I knew her, I pondered why she repeated this assurance like a mantra. In general, Um Mina was unusually persistent in showing and explaining to me her wisdom as a housewife and mother: her vigor in cleaning, her attention to the details of her children's education, the propriety with which her three daughters conducted themselves. But this refrain was tied into an anxiety about Ezba that I often encountered. "*Bitaʿrafī minninā?* [Do you get disgusted by us?]" was always the first question I was asked by women when I made a polite refusal when offered refreshment—part of the normal give-and-take of a hostess's duty to press a drink on guests and the guest's duty to politely refuse until sufficiently entreated.

Um Mina's concern over being recognized as clean indexes the social and affective repercussions of the material realities of water in Ezbet Khairallah, mirroring wider classist discourses nationally about urban suffering (Beinin 2010, 2001; Bayat and Denis 2000; Lockman 1994). Ezba's residents are sensitive to the messages of value coded in their area's variable access to official city services, and their bodies and homes are shaped by their enmeshment in these systems (Peteet 2017). These challenges focused around obtaining and distributing potable water and managing the constant presence of the smell of sewage as people moved in and out of Ezba's spaces, leaving impressions and forming relations. Residents of Ezbet Khairallah work within these systems of received possibility to articulate particular kinds of social relations and to manage shame around excess wastewater and attendant disgust.

Conclusion

Sowing connection has been a long-term, multistrand, and nonlinear process for the inhabitants of Ezbet Khairallah. As people moved to the area, they gradually claimed what had been a wilderness space, garbage dump, and unofficial cemetery, building homes and lives and connections to each other. Through social, material, and ethical labor, residents of Ezba rooted into the sandstone ridge on which the area is built and attempted to draw in the water and resources necessary to cultivate a flourishing community. Unsurprisingly, these efforts were opposed by a variety of formal institutions that wanted to pluck them as if they were weeds and to steal the seeds that they were planting and tending in order to build luxury housing or manage archaeological ruins on this contested land. The GOC's and MCDR's attempts to raze the settlement showed residents over and over that their presence was unwanted. They fought in many ways and at many scales to make their lives and their neighborhood a possibility. They rebuilt homes, they organized informal water and wastewater systems, they filed lawsuits and fought court cases, and they hauled water on heads and shared water with neighbors. Overall, they were the best urban farmers that they could be, adjusting to the uneven rhythms and possibilities of water. They cleaned houses, cooked food, visited neighbors, and built social relationships on, around, and through the material pipes and joints that carry water, hold water, separate water from sewage, and manage its overflow.

Water connections—both material connections in transport systems and social connections in the neighborhood—are central to the project of being well connected in Cairo. Through water, residents create the conditions needed to thrive. Inhabitants know that the GOC, the MCDR, and the Ministry of Antiquities think that they are getting in the way of the "real" value of their location. They know that they are targets for dispossession in service of capitalist accumulation, as the ISDF and the Cairo 2050 plan lay out (Tadamun 2015), and they continue to respond tenaciously by tending the roots of what they plant. This process of sowing is messy in many ways. Buildings with nonstandard heights and self-made piping structures must be managed, both materially and socially, and they are overlaid with ethical considerations for neighbors and deep attachments to place.

During the implementation of the formal sewage system, I watched my friend Um Amina, in a perfect embodiment of the neighborhood, negotiate precisely these tricky waters in speaking with her neighbor about his front doorstep. Um Amina loved to have me visit her small soap shop, as she said that my face was good for her business: the added curiosity helped draw women to come by. Her elderly male neighbor, whom she referred to respectfully as Shaykh Mahmood, regaled us with a long tale about the poor construction of his stairs. He described how poorly built the stairs to his house were and how they didn't go all the way to the street level. He explained that the builders had told him it was the only way to do it years ago when the house was constructed. Um Amina replied, "Well, I didn't say anything about your stairs, but yes, they are terrible." During this conversation, Mahmood also repeatedly told Um Amina that his wife was trying to get people together to fix the road and make it more level now that the sewage pipes were going in.

When he left, I asked her why the very detailed conversation about stairs was necessary. She laid out for me what Mahmood had not said directly. He was really trying to explain, she said, why the builders had put so much more dirt on their side of the road, which meant that water was always flowing down to her doorstep and filling her septic tank. The story of the poorly constructed stairs was really a story about unmanageable water, and his visit to her soap shop was a project to maintain good neighborly relations in the face of undesirable material constraints. Was she mad, I asked. "No," Um Amina replied. "At this point it was just forced on them [*ghaṣban ʿanhum*]."

Chapter Three

Locating Connection

Sitting on the ground cross-legged, peeling potatoes and snapping beans for dinner, Um Marwa turned her chat with neighbors to the tricky management of her eldest daughter's engagement. The late afternoon sun pushed its way through the cracks of wooden shutters, lighting the halo of hennaed hair above Um Marwa's expressive eyes: "We decided to have the engagement party here. Even though our apartment is small, and Ezbet Khairallah is not a nice place to bring visitors, we wanted to do the party here rather than in Sohag. My husband insisted." This last was by way of praise for her second husband's generosity. Um Marwa had been a young mother with two small children in Sohag when her first husband passed away. After a few years, she married her current husband and moved to Ezbet Khairallah in Cairo, where he has a small workshop, leaving her eldest children to be raised by her brother and his wife in Upper Egypt. Now her eldest daughter, Marwa, is in her late teens and engaged to a young man who has Um Marwa's conditional approval. The engagement party had been held for the young couple a few months prior, but it was a mixed success. "I wish you had been there to see it!" Um Marwa said to me.

Her neighbor Um May looked up with a laugh from her inspection of a bowl of raw rice for stray small stones and insect parts: "Why?! It was a disaster!"

"No," Um Marwa replied, "she would have been interested. For her research. But it was embarrassing to bring people to Ezba." Um Marwa repeated a sentiment that I heard often in Ezba: it is uncomfortable to bring guests to the area. How can you explain away sewage on the streets to prospective in-laws and country relatives? How can you invite guests to an area with a reputation for crime and danger? But Um Marwa's husband had made up his mind, and they had prepared for the party to celebrate the engagement. The arrangements for such affairs are always complex, negotiating who must be invited and who is likely to take offense, renting an elaborate colored gown for Marwa, the bride-to-be, choosing a makeup stylist, preparing the house and food to host relatives from out of town, and purchasing sweets and soda for the party guests.

The women then told me the story of the disastrous engagement party. A few days before it was to take place, Um Marwa had begun to make plans for how to manage the wastewater demands of her party. She needed to have the septic tank emptied in order to accommodate the extensive use of water that the party would entail, but she couldn't empty it too early for fear that her neighbors in the apartment building would have filled it up by the time of the event. As a solution, Um Marwa's husband paid a small bribe to the driver of an emptying truck to come on the morning of the party, and Um Marwa spoke with all of her neighbors in the building, who would be guests at the party, asking them not to use much water on that day.

The day of the party, however, the emptying truck driver failed to show. Um Marwa was in a quandary. Her dutiful spouse marched off, looking for another emptying truck. Neither luck nor a sufficient amount of extra cash being available, this plan failed. How was Um Marwa to manage a house full of guests without using water? And if she did use water, how was she to manage the disapproval of neighbors and passersby, as well as guests of the party, as her septic tank filled to overflowing? By late afternoon, Um Marwa told me, she had decided not to be conservative with the water use at the party. However, by that time, the question was mostly irrelevant. The tank had already overflowed, and the uncontained sewage bubbled up in a pool above the septic tank and ran down the street.

Um Marwa described how later that evening, the bride-to-be picked her way carefully down the street to where a stage and lights had been set up. She was a vision in a huge peach-colored hoop dress and matching veil, lifting her skirt out of harm's way. Um Marwa had placated as many neighbors along the street as possible by inviting them to participate in the festivities, and she later took small gifts of sweets to other neighbors who had not been invited, trying to soothe feelings over her perceived lack of consideration. These small gestures of connectivity and neighborliness were markers of sociality in the context of the porous and often overflowing septic tank system, which Ezba depended on for roughly thirty years. In this story, Um Marwa held relations at the threshold of rupture and tried to figure out how much pressure she could put on her community.

The preparation for Marwa's engagement party demonstrates the social labor involved in mitigating the effects of unruly sewage, a kind of sewage etiquette, which is required to maintain neighborly connections and more diffuse cordial relations in Um Marwa's local context. This is one aspect of a larger universe of sociality in which connections must be discovered, evaluated, and established. In this chapter I look at locating connection in two senses: first in the sense of identifying or finding, as people "locate" possible social connections and build social networks; and second in the sense of geographical and spatial positioning, for connections are mediated at physical thresholds, both at the doorsteps of homes and within homes. I show how connection indexes both processes of discovery and processes of instantiation.

At thresholds—the doorsteps of homes and businesses and under the roads in front of apartment buildings—residents mediate the public appearance of private waste. The threshold is a key node of connection between internal household arrangements and external infrastructures, as well as a central location of social connectivity among neighbors and larger social networks. I use the term "threshold" to refer to the places where the sociotechnical systems of infrastructure laid out in streets meet up with the internal networks that constitute domestic spaces. I also consider thresholds within the home, such as the bathroom, which is a particularly important location of public and private distinction.

Locating connection in the physical area of Ezba involves discernment, as residents work out where connection ought to or could happen and work to create or limit that connection. Connections depend on particular kinds of inside-outside and public-private distinctions, and the neighborhood's larger projects of instituting infrastructural systems rest on connections between political, social, economic, chemical, and biological actors (both sentient and nonsentient) at a variety of scales. Piped networks in the street must meet at (or under) the doorstep with internal networks that distribute water within homes and other buildings. These local projects are not necessarily coherent with one another, and the forces that drive them can pull in many different directions. Bending these multiple forces into convergence at thresholds is an ever-evolving project that helps the community to root where it sits, to embed itself in a particular location, which in this case is a marginal urban space where connection has long been fraught and continually subjected to processes of deliberate uncertainty.

Just as material realities of infrastructure meet at the doorstep, so do social connections, which residents maintain in order to craft and strengthen human ties that are about love, companionship, safety, and possibility. In this chapter I observe these social channels flexing, strengthening, or redrawing pathways in response to the piped sewage system. This new wastewater management system, with its different material connectivities, created new inflections in social networks that required additional investments of social labor. I therefore also locate connection in a set of questions that women ask about building social networks. Where can I find others who will help me? Where can I go for resources like money, time, and physical work? Who do I know who could make a task possible? Women can locate connections in many ways: by chance, through hard work, by overhearing information, or by making astute guesses about who is doing what and with whom. Women also ask questions about forging and mediating these connections once they are located. How do I build affection and mutuality with this person? How do I make sure that connecting *here* doesn't create a problem for connecting *there*? How do I mitigate the impact of the socially awkward public appearance of sewage when I have limited means? Um Marwa negotiated these questions in her connections around the engagement party. The invitations to the party, the social visits, the small gifts, the recognition of the inconvenience to others—all were labor invested in discerning where and how connections existed, whether they were impacted, and how to ameliorate that impact to preserve existing and prospective relationships.

Social infrastructure relies on practices of phatic labor, a form of noninstrumental labor that creates communication channels for the flow of sociality, information, and, potentially, resources (Elyachar 2010; see also Stamatopoulou-Robbins 2019; Simone 2004b). The work that women do to create friendships and maintain positive social relations in a larger social context is an indispensable component of sociotechnical systems, particularly in the urban margins, but this effort often goes unnoticed as labor. This type of networking labor is invisible or ignored for several reasons: because of artificial categorical distinctions between the technical and the social, because women's labor is undervalued, and because of the polysemy of the labor itself. (If done well, it is not merely instrumental, and it is never entirely outside a context of mutual dependence.)

In Ezba, this social labor is operationalized in the management of unruly sewage, which confounds a desired set of private-public distinctions. In this context, the work of locating requires women to arbitrate appropriate levels and moments of connection and disconnection. In Ezbet Khairallah, public and private are not normative, mutually exclusive absolutes; they have their own set of relational meanings. Bodily functions, such as bathing and showering, are private affairs that produce distinctions between spaces (inside and out, general and specific, public and private) and between people in social relation to one another (self, family, friends, neighbors, strangers). Spaces such as bathrooms, where bodies are naked and functions are smelly, generally are weighted with shame and managed with relative discretion (McGlotten and Webel 2016).

According to Dominique Leporte (2002), the smell of shit became a key marker of familial intimacy in the rise of modern cities. In Ezba, though, sewage refuses to remain hidden. It pushes up into public spaces outside homes, transforming the private biological functions that everyone tacitly agrees not to acknowledge into a visual, tactile, and olfactory public problem. This operates at the level of the home, opening a household or building to censure for sewage intrusions into the street, and at the scale of the neighborhood, as streets paved in sewage-fed mud entrench ideas about the abjection of the neighborhood (Anand 2012b). It matters when matter is out of place (Douglas 1966), particularly when that matter helps to elaborate social and environmental boundaries. In Ezba, women serve as the primary mediators of public and private intersections—functions that in more affluent areas of the city might be outsourced to a maid or a doorman (*bowab*).

Projects and pathways for locating connection deconstruct and reconnect in relation to shifting technical arrangements, and this work is undertaken primarily by women as they test, establish, and reestablish the locations of connections. Ezba's move from a system of independently constructed septic tanks to a state-sponsored sewage network follows the logic of technological progress, for it increases the informal "backward" area's incorporation into the utility embrace of the municipal government. However, as I observed the everyday processes of implementing the system in 2009 and 2010, it became clear that the new system would operate to reorganize the burdens of connection rather than to eliminate them and that urban residents of limited means would face financial and social obstacles in their pursuit of connecting well to the new order of "things" (Appadurai 1988).

I examine these changes through stories of sewage and sociality, ethnographic moments in which residents of the area attempted to locate connections in social networks and create connections at particular locations. First, I introduce key issues in wastewater management that set the context for conditions of possibility in Ezbet Khairallah. I describe everyday life in Ezba before the official sewer connection, highlighting the material and phatic labor that made life possible in the context of the septic tank system. I show that bathrooms are internal thresholds at which the mediation of private waste is enacted; they trouble any absolute distinction between private and public.

I then describe the municipal system intended to replace the septic tanks, focusing on the materiality of the doorstep as a key space for negotiating the putative end of the old system and the much-anticipated approach of the new piped system. This ethnographic moment centered on how one system was broken open to make way for the next, and I trace residents' social labor in connecting to the construction of the state system.

I also follow the fate of the septic tanks, which the advent of the piped system made obsolete. This story captures a moment in which the old spaces just beyond the doorstep made visible the reach and the limitations of neighborly assistance. In the final section of the chapter I again cross the threshold of the home, showing how the new wastewater system imperfectly intersected with the potable water system, tracing the culture of expectations for intimate mercies and the limits of financial possibility. Social systems change with infrastructural regimes, and the threshold is a key location for the entwining and evolution of material and social connections and the various possibilities for making those connections function well.

The Problems with Wastewater

Wastewater systems, the less glamorous sister of potable water networks, are central to human health and the successful functioning of cities. As Um Marwa's experience demonstrates, wastewater offers multisensory notifications of missed connections and poorly located relations, and it is therefore a good location to think through the stakes of connecting well. Service infrastructures expand to provide more spaces with access to potable water, and older waste and runoff management systems are easily overwhelmed by the new volume of wastewater, producing a need for new sanitary systems that can manage the going out of all that has come in (Cousins 2017; Ranganathan 2015; Gandy 2004; Kaika 2004). Yet globally, wastewater system implementation lags far behind the expansion of potable water, often with devastating effects on human well-being and the built environment. As of 2015, 2.4 billion people across the world lacked access to basic sanitation services, such as toilets or latrines; each day, over 2,000 children die due to preventable water- and sanitation-related diarrheal diseases (Centers for Disease Control and Prevention n.d.).

There is a gradient of systems considered acceptable as wastewater management. The World Health Organization and UNICEF Joint Monitoring Programme defines an improved sanitation situation as one in which there is a separation of human excreta from human contact. Systems that meet this definition include flushing or pour-to-flush systems of piped sewage, septic tanks or pit latrines, improved pit latrines with ventilation or pit latrines with slabs, and composting toilets. These systems are the top rung of the Joint Monitoring Programme's four-step "sanitation ladder," which has the bottommost rung as open defecation in fields, forests, and bodies of water. While the septic tank arrangement long in operation in Ezba technically falls on the top rung of "improved sanitation," the system's porousness and tendency to overflow meant that very often human excreta were not as hygienically separated from human contact as this category requires. In Egypt, the United Nations Development Programme estimated that 97 percent of urban residents and 93 percent of rural residents have access to improved sanitation (Albuquerque 2010), although there is considerable variation between rural and urban locations and within urban areas based on socioeconomic status. Looking at the implementation of Cairo's first modern sewage system between 1907 and 1915, Shehab Fakhry Ismail (2015) lays out the ways that the implementation was a joint program of technological and social engineering, a logic that can be seen again in the contemporary moment as residents of Ezbet Khairallah adjusted to the newly engineered disjunctures at their doorsteps.

In Cairo, both solid waste and wastewater have played significant roles in twentieth- and twenty-first-century political contestations, marking out a charged political terrain within Cairo's urban politics (Arefin 2019; Winegar 2012).[1] Scholarly attention to infrastructure has revealed the ways in which material systems designed to manage the flow of things (water pipes, roads, electrical wiring, media) are deeply interrelated with ideological projects and social arrangements. This body of work has laid out the poetic and political nature of infrastructure itself (Larkin 2013) and the constructed nature of water as a resource (Barnes 2014; Carse 2012; Linton 2010; Kaika 2004), often responding to larger discussions in science and technology studies that aim to reorient notions of human-material relations (Haraway 2007; Latour 2004). In the literature on infrastructure, particular attention has been paid to the impacts that utilities have on citizen-state relations: water technologies that appear to be modes of inclusion for previously disenfranchised populations can, in fact, be technomanagerial sleights of hand in which contentious political debates are transformed into questions of technological capacity (Stamatopoulou-Robbins 2019; Ballestero 2019a, 2019b; Von Schnitzler 2008). Such technologies can also function as mechanisms of increased state surveillance by enabling new lines of sight into daily practices (Von Schnitzler 2013). In city spaces, access to water can brutally demarcate social standing and highlight the instability of the extension of state recognition (Anand 2017; Björkman 2015; Gandy 2014).

Water illuminates the tight bond between material elements and what people do with them both in the everyday and as elaborated in formalized systems, such as political entities, legal structures, and economic schemes. As Ezbet Khairallah shows, locating these connecting concerns at thresholds enables the tracing of the social and material labor that residents invest in making water and wastewater systems connect well. Ezba also illuminates the character of the old self-installed networks. The porous nature of septic tanks (*transhāt*) allowed sewage to seep into the streets and often to infiltrate the joints and fissures of the clay and later plastic pipes that residents had installed to carry potable water. Residents thus had to continually work to determine the threshold between water and wastewater.

The Septic Tank System

The system of *transhāt* (pl.; sing. *transh*) that operated in Ezba was a highly porous waste management arrangement. Over the decades as Ezba developed, brick and concrete buildings replaced the stone and mud rooms built by the first settlers and were accompanied by septic tanks. These were large concrete boxes, roughly 1.5 meters square, sunk into the dirt street in front of each building. Unlined and often covered only with a movable stone or a plywood sheet, the septic tanks were a constant source of seepage and overflow. All water that went down the drains or toilets in a household was flushed by a simple system of gravity and water pressure through a central pipe near the top of the septic tank. Squat toilets, kitchen faucets, and drainage grates sent shower water, laundry runoff, water from cleaning floors, excrement, and bits of food through the PVC piping—usually self-installed or put in by unlicensed plumbers—that ran from upstairs apartments through the open area in the center of the building into the tank in front of the house.

Different housing units filled septic tanks at different rates. A single-family home may have filled its tank every week or two, while the tanks of newer, more

Runoff from an overflowing septic tank, 2010. Photo by author.

densely populated concrete apartment buildings may have only lasted a day (or less) before needing to be pumped. In this set of circumstances, sewage etiquette helped to mediate forced intimacies, such as during the engagement party. Residents worked through social systems to allocate each one's responsibility. Some residents attempted to restrict their use of water at the individual level, while others were forced into restrictions by landlords, who shut off water for the majority of the day or for several days in a row in order to reduce the frequency with which the septic tank had to be emptied. Even after residents self-installed water pipes, they found their water consumption limited because their septic tanks could only handle so much.

Residents paid for the septic tank pumping, and multifamily homes split the removal truck's bill. Woe betide any woman who was away when the tank was emptied. One young mother in Ezba described her frustration when she returned from an overnight trip to her mother's house to find that the septic tank had been emptied in the morning, and neighbors in the building had rushed so quickly to clean their homes, wash their laundry and scrub their rugs that the septic tank had already become full again by the time she arrived home in the afternoon. This young mother's experience highlights that this septic system geolocated women's labor, as her short-term absence meant that she stepped out of sync with the rhythms of sewage management under this regime. She recalled surveying a pile of her young daughter's dirty clothes, which would have to wait until the next time the septic tank was emptied, and went to her purse to get money for her share of the removal fee. There were no discounts for a missed opportunity. Maybe next time she would think twice about an overnight trip.

Sewage was long a defining feature of Ezba as a location. Because the septic tanks were porous and emptied at individual expense and because the roads were so dusty and likely to turn to mud, all water that appeared in the street was subject to negotiation and contestation by concerned parties. When septic tanks leaked or overflowed, as they invariably did, the sewage in streets invited many vectors of disease: insects, contact contamination, and the contamination of potable water. There was also the kinetic health danger of loosely covered pits. One of the first stories that I was told when I arrived was of the recent drowning death of a young child who had fallen into an uncovered tank. Such instances intensified the already-common public attention to others' septic tank practices. When someone caused an overflow, they were very likely to get a visit from angry neighbors or to have criticisms repeated to them by shop owners, who listened to and passed on the complaints of customers and passersby. When the tank was full, people had to forgo or delay household cleaning and used only minimal amounts of water in a bucket to perform ablutions for prayer, to wash their hands, or to prepare food.

This system of private responsibility for mitigating a public health hazard was not always successful. During the installation of the state's piped sewage removal system, the cost of waste removal more than doubled, from roughly EGP 30 to EGP 80 per visit in 2010. It was not always easy to get tanks emptied when necessary; homeowners and landlords collected the cell numbers of the independent operators of removal trucks or walked to areas where collection trucks congregated to try to get someone when the time came to empty their tank. In addition, the process of estimating how full the septic tank had become was an imprecise one. The resulting omnipresence of sewage in Ezba streets marked both bodies and social relations in the area.

The first time I met Asmaʾ, I was lost in Ezba and turned down a narrow side street, no wider than six feet across. Sewage puddles spanned almost from wall to wall. Sitting on her doorstep was a woman having a cup of afternoon tea. She told me that she had just spent her morning ankle-deep in the puddles, pushing the wastewater out of her small street and into the main thoroughfare. Her neighbor, hearing our conversation, came out to join us and noted that the water had changed color from yellow to black. She pointed at a spot in the middle of the street, where we could see it bubbling up from the septic tank below. Both neighbors had been trying for days to get a removal truck to come clean their respective tanks, and Asmaʾ told me that the only way to get a driver was through bribery. What was there to do? When the septic tank was full—as frequently as once a week for this house—they had to go in search of a septic removal truck and hope that they could get it to come before the whole street filled up. This week, Asmaʾ hadn't been lucky and had spent the day trying to manage the overflow.

I asked Asmaʾ if that kind of contact with sewage would harm her. "It already has," she replied, lifting first one leg and then the other, pulling back her long house dress (*gilabīyah*) to midcalf and offering her feet for my inspection. It was a gesture I became accustomed to, often offered in response to questions about health and the impact of sewage, or as punctuation in the course of a life history being told. A body part would be displayed—without visible injury but offered in the spirit of proof. *Here, see this. It tells my story. This foot, this arm, this cheek lives in Ezba.* Asmaʾ told me that she wanted me to take the water problems in Ezba and give them a loud voice, to make those responsible hear. She also said something I heard repeated a number of times in my first months of fieldwork: "Do you think you could live here, in these conditions? Do you really? Not a day. You couldn't live here a day. Where in Cairo is there a good place for people to live? Zamalek? Garden City? Fine, if you have the money. But to each what he is able. People live here who are poor. What can they do? They will live here."

As I learned, Ezba residents live there—they live well there—through processes of connecting well, just as those in the wealthier areas do as they seek life chances. Asma' was theorizing what connecting well looked like in her context, articulating the value of embeddedness, of remaining in a location. She also offered her own body as a site on which harm was written, even if there was no visible scar. Her body, as it lived in Ezba, was a location where one could see the ramifications of managing sewage in that place.

Mediating Boundaries

In late 2009, I rode into Ezba with an acquaintance, Amal, on a private bus from the Abu Ashraf market area. We bumped along the roads, which were rutted and viscous from sewage that had percolated up from nearby septic tanks and flowed downhill from higher areas in the neighborhood. The bus didn't stop in Ezba itself, and while trying to get off the slowly moving vehicle, my companion fell backward, right into the black broth of a sewage puddle stretching some ten meters wide and twenty meters long on the main street. Amal's hands and the back of her skirt were covered in the sewage mud. I helped her up, and we made our way to the nearest vendor, a small storefront selling *ful* (bean paste) sandwiches, but we found that there was no water connection in the store. The shopkeeper, who did not have sufficient water to help Amal, took an empty plastic bottle and went across the street to a *koshari* store to fill it. (*Koshari* is a popular Egyptian street food made of lentils, rice, macaroni, and onions and topped with tomato sauce.) The man gave the bottle of water to Amal and asked us to do the cleaning at a little distance from his store. We did the best we could, but with only a single bottle of water, the attempted cleaning only spread the waste further onto us both. I went to the *koshari* store and got a refill, but the damage to her skirt would not be fixed so easily.

We made our way back to the *ful* store to try to return the bottle, but the proprietor did not want it back. While we were standing there, a woman named Um Ameer, who had come to purchase things from his store, invited Amal back to her apartment to clean up. It was a kind gesture, bringing strangers across the threshold of the house, although other interior thresholds remained to be negotiated. I followed Amal and Um Ameer to the latter's apartment near the eastern edge of Ezba and was ushered into her guest sitting room, where I was left to wait. Um Ameer's apartment seemed quite ordinary for the area, with modest furnishings and cement walls covered imperfectly with green paint. In some quiet evaluation of Amal and me, carried out between the *ful* store and her home, Um Ameer had decided to bring Amal to her sister-in-law's apartment across the hall rather than take Amal into her own bathroom to clean up. When Amal returned, freshly clean but too wet to continue the day's planned activities, she reported that the apartment across the hall was really well furnished and quite lovely. Her assessment was that wealthy people lived there, which made it a preferable place to bring guests.[2]

Um Ameer asked us to join her and her children, a boy and a girl, who were eating their breakfast of *ful* and falafel sandwiches in the family sitting room. We declined, but I later returned to thank Um Ameer with a small gift, and we became friends. In that first meeting, I was struck both by the kindness of her invitation into her home and by her careful arbitration of our access to various areas within her home and that of her nearby relatives. By moving across thresholds—from the street outside to the interiors of private homes, from public rooms in the home to private bathrooms—one can see these boundaries as charged locations for mediating connection, which make visible differing levels of intimacy.

Sewage—both in the street and in the bathrooms where it is eliminated—indexes disgust, a potent affect that engages cultural notions of morality and embodiment. While the objects and contexts that are found disgusting vary across space and time, disgust itself has been recognized by anthropologists and social scientists as central to social relations (Bubandt 1998; Miller 1998; Douglas 1966). In particular, disgust is important to the drawing of boundaries between me and them, good and bad (Durham 2011), and between different groups of people (Manalansan 2015). Disgust has been used as a device in global public health education—often with detrimental effects (Brewis and Wutich 2019; Lupton 2015). Disgust is also politically effective in differentiating oneself from that which is too close to the self (Nussbaum 2006). And it has thus been a key affect in sovereignty struggles and embodied projects of difference-making (Winegar 2020). Here, I focus on bathrooms, which offer a view of the work that goes into managing the types and timing of connection and disconnection, the calibration of what it is to be well connected in different scenarios. Although bathrooms are thought of as individual spaces, they are also locations of connection as residents work to figure out who ought or ought not to be there, and how to manage the outflows of what happens inside.

Bathrooms and the toilets within them are an important location of social demarcation, differentiating both closeness and social status. Toilets are a key barrier to the transmission of diseases between people: fecal-oral disease transmission, "whereby pathogens in fecal matter from one host are introduced orally to another host" (McMichael 2018: 294). Water, sanitation, and hygiene interventions are part of active projects of national and international organizations, and these goals authorize a slew of NGO projects globally. For example, for many decades, NGOs and the state in India have sought to change sanitary habits in the nation, which has high rates of open defecation and limited toilet facilities. India has thus been the subject of a great deal of social science research on bathroom practices. O'Reilly and Louis (2014) distill the model of the "toilet tripod" from this research, which posits three broad forces that shape toilet adoption and use: proximate social pressure, or person-to-person interaction; multiscalar political will; and political ecology in the forms of changing land use, access to water, and soil types. As this theory would predict (and as Ezbet Khairallah's project of sewer system installation evidences), waste is both socially affective and effective (McGlotten and Webel 2016; Jewitt 2011), and it is always politically charged (McFarlane and Silver 2017; Redfield and Robins 2016).[3]

Ezba bathrooms came under close scrutiny as an exemplar of mismanaged urban waste in early 2009, when it was selected as a site of research on sanitation. In this study, Ezba was positioned as an "urban slum" and used as a comparison with village and peri-urban areas. The report, written by the Center for Development Services on behalf of the World Health Organization, examined women's and children's roles in sanitation. More specifically, the research interrogated the behavior of women and children around using public and private bathrooms, hand washing, and consuming water from communal sources outside the home. It examined these behaviors alongside viewership of media messages about proper sanitation behavior. The report was part of a larger market research project disguised as a public health campaign. The stated agenda items for this study were to find a baseline of current water and sanitation systems in poor areas in Egypt, to better understand gender and generational roles in managing water and wastewater, and, significantly, to assess the willingness of households to pay for potable water and wastewater services. Managing waste in appropriate ways is key to creating proper national subjects, but it is also necessary in the structurally adjusting state to assess these communities as future consumers of sanitation services. This report was also a justification for a continued interventionist policy, with inadequate waste management serving as a marker of lack and backwardness. As Stephanie

Tam (2013) says of colonial India, the state insisted that these residents lacked a sanitary civic sense. Or more colloquially, they didn't have their shit together.

Bathrooms are a charged space in Ezba homes. You need to know a family fairly well before you can ask to use the bathroom, and you need to do it in a pretty low-key way, especially if you need to defecate. Kitchens in many homes barely function, with blocked sinks and slow water taps, and when you only have water every few days, you often can't afford to wait for water to dribble out of the kitchen tap to fill containers, wash clothes, and so on. The heavy spigot in the bathroom thus often becomes the main water feature for the entire household. This is especially true for those living in apartments, where there is less control over the quality and repair of plumbing components. But even in freestanding houses, large water-related projects of all sorts often take place in the bathroom: cleaning clothing, washing out water storage bottles and tanks, bathing children, rinsing fruits and vegetables, and washing big piles of dishes. Figuring out how to divide tasks by space seems to be based on the volume of water needed or the amount of work to be performed, rather than on the nature of the task. A tray of teacups or a snack plate could be rinsed while standing at the kitchen sink, but the pots and pans from a large meal were more comfortably done in the bathroom, squatting or sitting on a low stool before a large tub. Elbows on thighs, legs straddling aluminum or plastic basins, with thin plastic flip flops on their feet to prevent rheumatism from traveling up from cold tile floors and settling in their hips, women in Ezba squat or perch on overturned crates or buckets, sinking into a posture of labor.

Where people can do so, they build sinks outside the bathroom proper, or in some other way they delimit the space of the sink from that of the toilet. Constructing this separation between toilet and sink, where possible, is one example of intentional disconnections made to prevent too much flow across the threshold between two worlds. This is necessary because toilet spaces are socially and morally charged on a number of fronts. In Ezba, bodily practices of eliminating waste were particularly sensitive, and simply entering this space meant moving from the public to the private areas of homes. In a house without a separation between toilet and sink, a guest might be brought to the bathroom to perform *wudu* [ablution] and have a fresh bar of soap opened in their honor, while the host or hostess waited near the door holding a towel for hand drying. Once a certain level of familiarity with a family was established, one could move into and out of the bathroom without being accompanied, but never without notice. A number of Ezba contacts told me that the bathroom is the place to really see if a woman keeps a clean home, so they are also charged sites in the moral economy of good home stewardship.

Bathrooms are also spiritually charged, for they are a prominent location of jinn (supernatural others), and there are very particular rules about bathroom behavior and speech, which are intended to maintain good relations with one's jinn neighbors. In the bathroom, all religious speech is considered inappropriate, particularly the mention of God. One should always give warning before throwing hot water, as hot water returns jinn to childhood; this is understood to be a painful experience, and it poses quite an issue for beings with such long life spans. Turning a jinn back into a child and other unneighborly bathroom behaviors are sure to earn you the ire of the jinn's family. The jinn could possess you and make you sick and crazy, which is already understood to be an issue in Ezba because it was built on a graveyard; some people there already feel themselves to be risking too close a relationship with spirits and jinn.

The fraughtness of bathroom spaces is clear in Um Wasam's home. Um Wasam lives in an apartment block near the top of the plateau on which Ezba rests. This is the area in which many of the new concrete apartment buildings have been built; it is located near the one government-run elementary school within Ezba's boundaries and is situated on the main market thoroughfare. Um Wasam's three-bedroom apartment had a single bathroom, fairly large by the standards of Ezba. The sink inside had been broken for months and had become a repository for the dirty linens of the youngest child. There was a flush toilet, though the flushing mechanism no longer operated, and it required a large amount of water to be poured into the bowl in order to force waste down the pipe. When the shared septic tank for the apartment building was full, waste simply accumulated until an opportunity arose to flush it, or the waste became too much to bear. Um Wasam's toilet was also the receptacle for some kitchen waste fluids, such as frying oil that had been used often enough or that had been used to fry fish. Like all women in Ezba, Um Wasam kept a series of plastic and aluminum tanks, barrels, and jugs to store water for bathroom and washing purposes. In the corner of the bathroom near the drain, she had a large blue water barrel set up on a bucket; a tap had been built in near the bottom of the barrel. This, she said, was to avoid contaminating the water when using a dipping cup—a problem with the open aluminum containers. The lid to the barrel was a wooden plank, and on the lid was a pile of children's clothing awaiting washing. In another corner sat a *baladi* [local] washing machine, an open drum with an oscillating paddle.

I was a frequent visitor in Um Wasam's apartment and had spent time in the bathroom with her, cleaning out water bottles when the landlord had unexpectedly turned on the water supply. I therefore felt fairly comfortable entering her bathroom. On one visit, however, I entered the room as she was using a squeegee to clean the floor after one of the children had not been able to wait her turn for the toilet. Startled and embarrassed, Um Wasam told me to get out, to go and sit in the guest parlor, a room I hadn't entered for months because I was no longer treated with such formality. A short while later, she joined me in the parlor, and for the rest of the visit, she studiously avoided any mention of bathrooms or bodily functions. The ubiquity of sewage on the street had not erased the taboo around bodily excretions, and the moment was embarrassing for us both. Indeed, the project of accurately understanding and navigating bathroom etiquette is an often underappreciated factor in anthropological fieldwork (Frøystad 2018).

Transitioning to the Piped Sewage System

After decades of self-managed sewage and the attendant health impacts of sewage overflow, residents in Ezbet Khairallah began working in earnest to change the sanitation situation there. In daily complaints to the Cairo Water Company office and in mass public actions, such as the blue jerrican demonstrations in 2008, residents of Ezbet Khairallah demanded that the Governorate of Cairo respond to their wastewater problems by installing the necessary piped infrastructure to carry away the sewage that lingered in place. Wastewater overflow from the area was increasingly pressing: it moved down the hill from Ezba and "menaced" nearby historic sites, and it infiltrated the porous limestone on which the area rests, increasing the risk of rockslides, such as the one in el-Duwayqa. Political pressure built along with the wastewater pressure, erupting into public view as residents took to the streets to protest. In response, the Cairo and Alexandria Potable Water Organization (CAPWO) and the GOC agreed to install a sewage system in Ezbet Khairallah.[4] However, given the long history of contestation over the location and permanence of the area (see chapter 2), residents sometimes found it difficult to believe that the changes were being made on their behalf.

The new sewage system was being installed throughout my 2009–2011 fieldwork. The workers began by uprooting the main streets of Ezbet Khairallah, digging long, wide, deep trenches to hold the main pipes from which smaller branch pipes would extend. In several areas—along the western ridge of the plateau near Zahraa, on the main road into the area on the eastern side of Ezba, and along the key thoroughfare of el-Negh Street—excavators burrowed deep into the compressed layers of dirt and garbage, opening grooves like the indentations of large fingers in the dough of the earth. Concrete pipes and other supplies were stored along the eastern road into Ezba and piled in small pyramids near the small mobile building that served as an office for the contractors. Later, as the project expanded across the area, trenches in the smaller streets were dug by hand with pickaxes and shovels since the large machines wouldn't fit. Throughout the project, workers had to avoid the buried septic tanks and self-installed water pipes. Hajja Mona, whom I mentioned earlier, pointed to the overlapping nature of the self-installed utility infrastructure: "In the time of the septic tanks, you couldn't smell clean water [over the scent of sewage]. There was no organization. The potable water pipes were right next to or laid across the septic tank. The electricity cables were across the septic tank. The world was all mixed up." In Ezba, digging was a messy and hazardous business.

The deep trenches on the main roads choked the traffic, making it difficult to cross the thresholds connecting the different areas of Ezba. Alternative routes for the main streets were hard to come by, and the construction work cut the connections into and out of the area, dividing it into eastern and western segments. The project affected pedestrian travel too. Along the main roads, earth was piled up along the sides of the furrows, great heaps that became compacted as residents made passages in the dirt. Women loaded down with small plastic bags full of daily produce, men on cell phones, mothers carrying children straddling one shoulder—all picked a careful passage along the edges of these deep drops. But stores in the market area of el-Negh remained open while that street was upended, and old wooden planks became temporary bridges at irregular intervals so that people could cross from one side to the other.

As the pipes in the main streets were completed and the smaller branch connections were put in place, residents began to connect their homes to the new system. Like others in the area, Um Amina and her husband hired the young men working on the main system to make those connections. The men worked on these individual connections in the evening after they had completed their arduous labor on the state-funded pipes. A labor subcontractor had brought these young men in from Upper Egypt, where many of Ezba's families had come from, and residents felt a lot of sympathy for them. People offered the men cigarettes, soda, and other small gestures of kindness and connection as they labored in the streets in front of residents' homes.

Connecting single-family homes to the new system often required at least the bottom floor of the building to be broken open and replumbed, and it often involved a process of renewing pipes and connections throughout the building. Several residents used the opportunity to renew and retile their bathrooms, often in a ubiquitous pink Cleopatra-brand tile. Um Amina invited me to see the state of her house as they put in the new system. The projects of deconstruction and reconstruction overlapped in the period when the new and old systems coexisted. At that time, the center of Um Amina's small street was open, and workers spent the day hand digging the trench to lay the pipe for sewage discharge. The section directly in front of her house had been opened at the same time, and the same young men came in the evening to rework the house's system and connect it to the newly laid pipes outside. Laying the connection from the doorstep to the branch pipe in the center of the small street was the responsibility of homeowners and landlords, who had to both arrange for the labor and materials and pay for them. The larger process had cost Um Amina's family several thousand pounds to date.

Inside Um Amina's front door, the floor of the small foyer had been opened by brute force, and piles of broken tile and chunks of concrete had been moved out of the way in a line from the door to the *manwar* (the small central shaft of the building where pipes run between floors). There was an additional short detour to the left where a pipe had been laid to a downstairs bathroom. The PVC pipes installed some twenty years before had been pulled up and cast aside, and new PVC pipes (reportedly of better quality) were already partially in place. During that week of disruption, Um Amina said that the household was not using any running water because there was nowhere for it to go. They went to a neighbor's house down the street to use the bathroom, and the rest of the household's needs were filled by conserved water—drinking water stored in reused soda bottles and grey water held in old plastic buckets until it could be dumped in a sympathetic neighbor's drain. When I returned a week later, the floor had been returned to normal and covered with new tile, and the connection outside her house had been buried. The center of the road near her, with the new state pipes, had been closed as the work continued on the next small street.

Um Amina's threshold was de- and rearticulated, highlighting the material ways that her house is connected—both outwardly to the state pipes and inwardly to interior spaces. The physical labor of connection was completed at night through an informal arrangement with the young laborers, who filled in a crucial disconnect in the governorate's material and financial plan: connecting the state pipes running down the center of streets to the pipes in homes. This connection was facilitated by the social labor of Um Amina and other women, who made tea and offered other small gestures to the laborers during the day. These small kindnesses eventually made possible the paid but informal relations that would connect their homes to the sewer system. The interactions also often incorporated idioms of fictive kinship with the young men, who came from the same region of the country as many of the residents. However, not everyone in Ezba could afford to pay laborers to make those connections, nor were there enough skilled men to connect each house in the evenings after the daily work was done.

Out with the Old (Tanks), In with the New (Pipes)

As Um Amina's story demonstrates, homeowners had to personally mediate the transition from one sewage arrangement to another, and this work often fell on the women in the household. This type of labor is familiar to women, who must constantly ask where they can find connections, who can make material connections and disconnections work well, and what social labor they need to invest to make those projects possible beforehand and continue afterward. Near the end of my fieldwork, the street where Asmaʾ lived came online with the state sewage system. I visited her while she was having her old septic tank drained. Two men were there with long cloth hoses and a simple pump, bringing the water up from the septic tank and putting it in the newly installed septic system in the main street. As I arrived, they were pulling the hose up and preparing to leave. Asmaʾ was telling them to keep the end down there until all of the water was gone. They were insisting that there was very little water left in the tank and that the pump motor would burn if they left it to work when the water couldn't reach the intake filter on the hose. She took several rocks and threw them forcefully down into the septic tank, looking to hear telltale sounds that would demonstrate how much water remained. Here, there were different notions about the status of connection and where things may go wrong. The men articulated a technical mismatch; Asmaʾ read it as an interpersonal mislocation, and she used social location to shame them into action.

Finally, the younger man got the older one to appease her by lowering the hose again and running the pump briefly. They did get a little water to come up, but not much. Asmaʾ continued to argue with and insult them. This provoked one of the men to pick up a brick and threaten her with it, although without much intensity. I came to sit next to her, not sure how the situation would unfold. They asked about me, and she told them that I was from America and had come to see the people of Egypt, the good people, not the smelly sort that they were. "Shouldn't you be ashamed to act that way in front of her?" she asked. The older man said

Replacing the old system with the new, 2010. Photo by author.

that he would further clean the tank *ʿashān khāṭrik* (as a special favor to her), which made Asmaʾ laugh: "As a favor to an American, they would finish!"

When it was clear that not much more water would come up, they unplugged the motor from inside her house and pulled the hose back up. Asmaʾ told them that she wouldn't pay them if they didn't clean the entire septic tank. They told her that it couldn't be cleaned more than that, and she said, "As you like. I won't pay you if you don't clean it completely." As the men started winding up the hose, trying to empty the water inside it into the new wastewater system, they ended up draining all of it back into her septic tank. Meanwhile, a neighbor's six-year-old son was enjoying being in on the action. He kept trying to help get the water out of the hose, although the men and eventually Asmaʾ yelled at him to get out of the way. Finally, the workers put the pump and the piled hose on a pushcart and dragged it away under a barrage of insults from Asmaʾ.

Now, Asmaʾ began looking for a social connection, a support for her position, someone to arbitrate in the dispute. She got a neighbor to come and look at the septic tank to tell her how much water was still in it. The neighbor told her that there was just a little water left, showing her where it would reach if his hand were in it—from his fingertips to the middle of his arm between wrist and elbow. She disagreed with him, saying that half was still in there. He said, "This is what I see. You asked me to look. I looked and told you what I thought. If you don't want my opinion, then it's none of my business," and he returned to his home. In this case, the connection didn't support her position in the way that she had hoped.

Asmaʾ then started putting bricks and debris into the septic tank to fill it in. Unsure of whether she was trying to eliminate piles of unwanted materials around her house or fill in a dangerous place, I joined her and helped move some discarded building materials from the street. She got garbage from inside her house and from her in-progress bathroom renovation, working to fill in this now defunct location. Then she tried to put the cement cover on top of the old septic tank. It was much too heavy for her, so she sent a child to find a male neighbor to move it for her. In a series of mislocations, several refused, citing work, or just didn't come, but she kept trying until she eventually found one who would help her. While we were waiting, the younger pump operator yelled from down the street that she should not leave the septic tank without a cover because a child could fall down into it. She told him to go away and mind his own business.

A little while later, the same young worker came back and chatted with her, telling her she owed him EGP 60. To my surprise, the man turned out to be her neighbor. Connections aren't always what they seem at first glance, and intimate connections such as neighborliness don't guarantee smooth operations. Asma' said, "Not just my neighbor but a friend, like a brother" (showing two index fingers entwined). "Ah," I said, "I could tell that you love each other from the way that you speak to each other!" They laughed, and he said that he was used to Asma' by now. She still refused to pay him, though a few days later she reported that she did eventually give him the money.

In this case, the material vestiges of the septic tank system called forth social labor from area residents, who engaged the existing social infrastructure that they had accumulated over time in attempts to manage new disjunctures. Asma', despite the conflict involved, was able to fairly successfully bridge the material gap by activating her social infrastructure. In the next story of sewage and sociality, though, Ezba residents struggle with multiple misaligned systems and struggle to connect well enough to manage the cost of the municipal sewer connection arriving at their doorsteps.

Water Is Heavy

Progress in Ezba has long followed a kind of rough teeter-totter, as improvements to one aspect of the infrastructure make gaps in related systems all the more obvious. The sewage system coming online carried the promise of round-the-clock water, but it ultimately defaulted on this promise because authorities failed to imagine the multiplicity of locations where connections were needed. While visiting with residents of one apartment building, I noted how excited many were. The previous day, the apartment building had finally gone online with the municipal sewage system, a major step forward that would finally allow residents to use the building's robust self-installed potable water system since they now had a place for all of the water to go. Under the old septic tank system, the landlord had severely limited the hours of water access for the building (which capped the amount of rent he could charge), and even when the water was on, it rarely reached above the second floor of the seven-story building because the single pump was too weak. Women on the upper floors had to come down to their neighbors' apartments while the water was running, fill buckets and jerricans, and carry them up the stairs on their heads.

Now that the sewer pipes were online, the building could theoretically have unlimited water in every apartment. But there was still a gap in the system: getting the potable water from the subterranean pipes to the top of a seven-story apartment building, which required strong pumps. The landlord gathered the residents and told them that he needed EGP 100 from each household to buy two new pumps to get water all the way to the top floor of each side of the building. People on the first and second floors, including Um Wasam, did not want to pay because they were getting water with the existing pump. Returning from work, Madam Nafysa found everyone gathered at the entrance to the building. An elderly woman from the fourth floor was yelling at Um Wasam that she must pay. Um Wasam was yelling that she wouldn't. How could she spend EGP 100 that her husband had earned with the sweat of his brow on water for other people?

The conflict unresolved, the landlord told them that he would run the water from 6:00 a.m. to 9:00 a.m. daily until everyone got their money together to pay for the pumps. Neighbors stated that it was a shame that with the sewage system running, there was less mercy now from both the landlord and neighbors than there had been before. It should be the opposite, they argued, especially since Ramadan was going to start in a few weeks. I sat with Nafysa and her college-age daughter, Sumya, in their fourth-floor apartment as they complained about this lack of kindness between people:

> NAFYSA: At first our problem was the sewage, because it was broken. But now, that's ended. The sewage has been fixed. There should be mercy, but there isn't any. Those in control of the water have closed it. They don't want to open it up for anyone until they collect the money. They want 100 Egyptian pounds from each apartment even if they only have 50 Egyptian pounds. There are people who have [money], and people who don't have [money]. How will they [who don't have it] give money? We have one lady upstairs who doesn't have work. How will she pay? It's up to God to find her the 100 Egyptian pounds. And the lady who lives next to her, her husband died, for example, and she's raising kids. How will she pay? Is this not *ḥarām* [religiously forbidden]? This is the center of the problem.

Nafysa was angry not only at the other residents' lack of willingness to act collectively, but also at the landlord's expectation that residents pay for the pumps.

> NAFYSA: No one can live without water, right or wrong?

> TESSA FARMER: Right.

> SUMYA: Especially in Ramadan. During Ramadan, the water is always on. Everyone has guests, for example. There are dishes to be washed. The rugs need to be cleaned. The water needs to be running. Being without water is impossible. *Ḥarām*. God will keep account of their actions and punish them. We just

want a little water but no, some people don't want us to have what they have. There is one old woman on the fifth floor who has to haul water from the ground floor all the way up on her head. My neck hurts me too from carrying water. Water is heavy.

Sumya's frustration spread from the landlord to the neighbors below, who didn't want others to have what they have, despite daily seeing those who were suffering from the lack. "Shouldn't there be mercy for those who can't pay?" she asked, especially in the month dedicated to reminding those with bodily comforts of the deprivations of those who have less. Where was the kindness? Interestingly, the water came on while I was there that afternoon, despite the landlord's decree—which he intended as a hardship—that water would only run 6:00–9:00 a.m.

After hearing Nafysa and Sumya's version, I went down to sit with Um Wasam in her second-floor apartment. She did not specifically mention the fight, but she spoke about how difficult it had been to make ends meet. If rent was EGP 350, and other bills for utilities were maybe EGP 150, and her husband was in a small savings association to help pay back his brother's debt for EGP 500 per month, and he earned EGP 1,000 a month, what did that leave her to live on? What would she use to buy children's clothes and to feed them? What about school costs?

UM WASAM: But we are almost done.

TESSA FARMER: It's almost at an end?

UM WASAM: Yes, there are seven people left to pay in the savings association, so that means that we have seven more months of this.

Um Wasam's family were in such dire economic straits because they were working to pay off the debt of her husband's elder brother, who was living in their hometown. His creditor was going to send him to jail, so the younger brothers had stepped in to pay. Since they didn't have much money either, paying the debt put significant strain on the lives of their families. The financial insecurity left Um on edge, she said, and made her frustrated with neighbors who insisted she must pay for the pumps. The pumps were a good idea for her too, she pointed out: "Who wants to get up at six in the morning to catch the water while it's on?" If the pumps were installed, she would have all-day access to water every day. But where would that EGP 100 come from? Here again, the anticipated benefits of the state sewage system created new disjunctures that Um Wasam had trouble bridging, given her limited resources. In addition to the material and economic strains that the transition exposed, she struggled with the ruptures in the social infrastructure on which she relies for survival. Her questions reflected the project faced by all the residents: locating connections in their social network to help manage the newly reformulated material connection waiting at the threshold of their apartment building.

The location of the water itself matters, as the ease of procurement is mediated by both social and physical locations. While penthouse living in wealthier enclaves may index the older sense of being well connected, in Ezba living higher up on the hill or higher in an apartment building can be a real problem. After all, water is heavy, and verticality matters, particularly for those who can't count on mechanical means to ensure its timely arrival in the right location at the right time. Stacked living offers potentials for intimacy through colocation, but as both Um Wasam and Nafysa shared with me, there are limits to the possibilities of connection.

Location, Location, Location

The site of Ezbet Khairallah is desirable in many ways: it is centrally located, offers some respite from air pollution, and provides views of the Nile and other important scenery due to the limestone outcropping on which it is built. However, Ezbet Khairallah's location also is marked by urban exclusion, and its verticality can pose some heavy problems. In this chapter I followed residents as they worked to discover, evaluate, and potentially establish lasting social networks and material links in wastewater infrastructure. I traced the intersecting flows of changing material systems and the resulting reorganization of social systems, which produce and are produced by the technical systems that they engage. My focus on projects of locating connection made visible the labor that residents of the urban margins performed as they sought to bend multidirectional forces into convergence at the entrance to their home (as with Um Amina) and to mediate relations of intimacy within their homes (as with the bathrooms of Um Wasam and Um Ameer). By locating the project of being well connected at thresholds, I brought into view the social labor, moral precepts, and sensory frameworks that residents of Ezbet Khairallah drew on as they engaged with wastewater infrastructure. From throwing garbage into newly defunct septic tanks, to managing a party's wastewater needs, to worrying over the cost of a pump, residents acted to connect their interests to and to differentiate them from each other's and those of the state.

The threshold is a material space and a set of social relations that exceed it. Locating connection includes everyday practices of neighborliness and the cultivation of cordial social relations. These actions, undertaken primarily by women, make life possible in the historical, economic, and material contexts of their lives. The older arrangement of septic tanks that operated in Ezbet Khairallah placed particular kinds of burdens on residents of the area, but multistranded relationships built up over time allowed people to manage life in the context of that system. While infrastructure as a manifestation of the teleology of technological progress and the expansion of governmental services is profoundly impactful in creating the possibilities and foundation upon which social life is formed, those with limited resources can find it difficult to bring that "progress" inside their homes. The ethnographic moments related here highlight the phatic labor that was involved in locating connections; this social labor is not itself instrumental, but it creates channels through which instrumental help can flow. The project of locating connection involves both discovery and instantiation. Once made, connections must be maintained, and they are often reconsidered, with participants constantly measuring their desirability, practicability, and immediacy.

Projects that seek connection can remain incomplete, which is evident in the gap between the state-financed systems and the piped systems in homes. Once located, the connections' material instantiations can highlight social gaps, as with Um Wasam's financial quandary. Inside the home, locating connection involves discernment. In residents' decisions about who may use their bathroom, for example, they work out spatially where, when, and with whom connection is joined and when it is better to disconnect.

The state's project itself—the decision to connect this informal area to city services—is equally complex. The GOC's extension of the sewer infrastructure to Ezba appears to contradict the story of the steep withdrawal of state welfare functions in neoliberal Egypt circa 2009–2010. The sewer narrative is not one of state abandonment. But neither is it one of triumph, given the difficulties that residents faced in connecting the new systems to their homes. Instead, this multilayered story shows how social systems shift in the context of changing infrastructural regimes.

The stakes for the management and location of sewage are serious, for sewage is connected to both human health and the narrative of social location; it affects the well-being of neighborhood residents in fundamental ways. In Ezbet Khairallah, sewage was, in some strange ways, protective. As the sewage disappeared—and with it the physical and social stigmas that limited the number of people seeking to live there—Ezba began to transform. Since 2011, Ezba's population density and cost of living have increased dramatically. Many new people have come to the area, and rising prices ripple through the social networks that residents had long strived to locate and instantiate. Ultimately, what the sewers of Ezba show is that connecting well is not only about the infrastructure itself, but about the many ways that people negotiate entanglement across social and material thresholds.

Chapter Four

Hedging Connection

After a long trip to the market, a small group of women and I sat in the cool first-floor apartment of Teta (Grandmother), the affectionate nickname of a family's matriarch. Teta could rightly claim the title of *hajja*, having completed the pilgrimage while spending several months with her son, who was working in Saudi Arabia. Rolling miniature eggplants a few at a time from their plastic bag to cut the stems and hollow them out for stuffed vegetables, the ladies were in a joking mood. Negla', Um Sara's youngest daughter, took up her customary role as the teller of jokes, including one of the many Juha jokes. Once Juha went to take a train home and brought his donkey with him. He asked the ticket seller, "How much is the ticket?" The vendor replied, "One pound for you, half a pound for your donkey." "Okay," Juha replied, "give me two tickets for donkeys."

Juha is an Arabic literary and cultural figure who often plays the fool to comic effect, but his wisdom comes out in subtle ways.[1] The Juha jokes follow a certain template, where the wise idiot turns common logic on its head. In the joke told by Negla', Juha played the fool to save money, which is one of many ways to get just a little further ahead in each interaction. Fifty piasters saved on a ticket are fifty more piasters in the pocket for cigarettes or to save for long-term projects. Of course, the joke is not realistic—a train ticket had long cost far more than one pound—but Juha's crafty ability to save is appreciated by those who also work diligently to make each transaction take them just a little further. The joke hinges on *isti'bāt* (deliberate misunderstanding), which is a technique to manage the appearance of knowing or understanding something. *Isti*ʿ*bāt* is key to being a smart individual, to managing life well, and it is an example of practices to hedge connection. It is also significant that Juha is willing to make himself equivalent to his donkey in order to save money. The donkey, Juha's companion, represents the idiot who does what others are not willing to do. Like Juha, the donkey has a double nature. If you want to call someone stupid or graceless, you call them a donkey (*ḥumār*), but it is the donkey who has the special ability to see jinn.

A strong knock on the outside door to the building was answered by one of the children in the house. He came back and told the group that the man with the electricity bill was at the door and needed to speak to someone. When Um Sara returned to the group several minutes later, she was exasperated about the high cost of the bill. Entering into the topic and the joking spirit of the moment, Asmaʾ told a story of a recent interaction with a water meter reader. When he stepped into the front entranceway of her building where the device is located, the reader mentioned that he was thirsty and wanted some tea. It was an unusual request in the situation, but Asmaʾ said that she had indulged him and asked how he liked it: "Do you take your tea with a little sugar or just right [*shāyak khafīf wa-lā-maẓbūṭ*]?"

"No," he replied, "I like dry tea [*ana baḥibb al-shāy al-nāshif*]."

Not comprehending the subtext, Asmaʾ said that she smiled and went to the kitchen to boil water. He finished reading the meter and left without drinking the tea. It wasn't until later, she said, that she realized he had been asking for a bribe to read the meter in her favor. "Dry tea!" she laughed. Her laughter was self-deprecating and in appreciation of the resourcefulness of the meter reader. She was mocking herself for having missed his cue: he was opening the door to a negotiation for mutual benefit out of sight of the official system.

Being quick on the uptake in these sorts of situations—looking for even small ways to get ahead, like Juha—makes it possible to survive on the difficult margins of the city. Small acts of evasion, selective misrecognition, and cleverness (*shaṭāra*) open opportunities to earn or save a little cash. These practices are central to the project of what I call "hedging connection," in which residents deploy limiting, restraining, or diverting actions in their quest to make connections work well for themselves, even when those connections are double-edged. For example, since Asmaʾ did not follow through on the meter reader's veiled request, it is hard to say whether the savings in the water bill would have offset the cost of the bribe, whether she would have come out ahead. But the meter reader's action opened a gap, a chance to make the infrastructure of water provision bend in her favor just a bit. The laughter of Asmaʾ at this story is in part an appreciation for the inventive use of the idiom of hospitality, which both masked a direct request for a bribe and shifted the footing of the interaction to one of social obligation and a measure of kindness through tea.

Hedging connection is essential to life on the urban margins. People are always trying to answer a fundamental question. How do I get the most out of the social, administrative, and material connections that I build while letting them do the least damage to me? As residents of Ezbet Khairallah engage in systems that offer both benefit and harm, they attempt to produce equivocality in their own positions and to unravel projects of deliberate obfuscation by relatives, neighbors, and representatives of state institutions. Inhabitants deploy hedging in order to protect themselves in the very same connections that they laboriously build to secure themselves; it's about stretching without rupture in uncertain conditions. People work to go to the brink but not go over; they play chicken. There is a weapons-of-the-weak element to hedging (Scott 1998), in that people are doing things to evade power while not appearing to evade power. But hedging is deployed not only in relationship to state systems. In Ezbet Khairallah, hedging is also used in interpersonal relations because social ties are at the center of how people make things work for themselves.

Hedging might be done through playing dumb or selectively misrecognizing (*istiʿbāt*), by delaying (*yamaththul*), by sending someone in another direction with a small lie (*ṣābūna*, which literally means "soap"), or by changing the frame. In Ezbet Khairallah, discussing who was playing *istiʿbāt* or giving the *ṣābūn* was part of everyday conversations between individuals, and people constantly worked to parse the truthfulness of statements made by others. Residents knew that others told them things that might or might not be true, just as they knew that they could do the same in their own communications. In conversations, both speaker and listener were aware that the real meaning might not lie in the direct statement; the real meaning might be deliberately obscured or be discernible only contextually. In contrast to an overt lie, in which the speaker's intent is to deceive, in hedging the key is in the fact that listeners don't have the expectation of perfect fidelity of meaning to content. This is not to say that all statements were double in this way, but that both speaker and listener were always aware that they might be. Moreover, such practices in Ezbet Khairallah were not always verbal. Body language and gestures were among the many tools in the arsenal of hedging, and the proper reading of them was required in order to recognize and properly interpret it.

Lisa Wedeen ([1999] 2015) describes practices of acting "as if" in the political realm in Syria, arguing that these behaviors demonstrate the capacity of the government of Hafez al-Assad to enforce conformity. In her reading, it is the appreciable difference between what the state demands that one says and what it is likely that people think that demonstrates the al-Assad regime's ability to enforce discursive compliance. She argues that this same dynamic operates in the everyday, where it became a project of communal self-policing with its own set of counterdiscourses.[2] In the case of hedging rather than the surety of the lie (as with Wedeen), it is the ambiguity around the truth that creates the possibility for people to "get by" each other. The effort in Ezba was to produce uncertainty, because opacity is necessary for managing social obligations when resources are limited and there are expectations of mutual obligation. Hedging connection is an attempt to maintain relationships without necessarily exchanging resources, information, or certainty, which may be too scarce to be spared or may pose some potential danger.

Because hedging is iterative and contextual, cues for interpretation can be missed or ignored, as with the story of the meter reader. The fact that Asmaʾ missed the opportunity he offered pushes back against a lingering structural-functional reading of culture—an assumption that culture is a machine that just runs. This type of reading would figure that because bribery is a known form of exchange in Egypt, all Egyptians know the cues that invite or offer bribes. In fact, people are not all equally well adapted to their cultural forms, and learning to hedge connection *well* is a skill that one develops over time. People's attempts are viewed as more or less credible. Their facility in knowing when to hedge, what version of a wink to use, and with whom establishes a kind of viscosity, a thickness to their identity's fluidity, weighting how it shifts in response to context. Too frequent use of the *ṣābūna* will get you the reputation of *shaṭāra* (being wily), while being a poor interpreter can make you seem gullible or, conversely, untrusting and antisocial.

These multilayered systems of association and evasion, realization and inaction, connection and misdirection are means by which residents strive to produce and eliminate uncertainty in the social, administrative, and technical systems that make up their lives. Residents of Ezbet Khairallah operate under a local political and social theory about how to read what is happening to them—how to interpret communications from other folks in their local context as well as from state and state-related actors.

In this chapter, I illustrate how these mechanisms operate in the context of payment and water provision in Ezbet Khairallah by tracing how the state uses hedging to delimit residents' claims-making and how residents use these tools to manage both the uncertainties of recognition by the state and their intracommunity social networks. Hedging is, of course, not limited to payment issues, but the pressures that make hedging crucial are fundamental to interactions around payment both to the state and between neighbors.

I follow two discursive structures that speak past each other about how and why water pricing in Egypt is not working. One of these discourses comes from state and state-related actors and focuses on infrastructural legibility and water theft. The other comes from residents and refuses the premises of the state's framework, instead focusing on the arbitrariness of the billing process, which, in the eyes of Ezba residents, challenges the legitimacy of the system. I explore how projects of hedging connection change the frame (McKee 2016) of the discussions around water and water pricing, thereby shifting the moral terrain of water provisioning. State representatives use their power, backed with the state's power of violence, to frame local practices of hedging as lying and theft.

In intracommunity contexts, hedging is a permeating communicative channel through which people make known their needs or obscure their resources by producing ambiguity—by managing the *appearance* of knowing or not knowing things, of doing or not doing things, mediating uncomfortable connections through refusal and misdirection. I highlight several encounters between community members during the installation of the sewage system in which residents attempted to obfuscate the realities of their financial and material situations (e.g., when did people get water meters, what did they pay for water). While in many of these interactions it was not clear that others knowing particular things would impact someone negatively, it apparently was better to safely obscure than to be sorry.

Municipal Utility Embrace

As I have shown in previous chapters, the delivery of potable drinking water in Egypt has been plagued by issues of access, quality, and cost. For many decades, Egyptian systems for pricing water have resulted in charges that were too low to cover the full cost of pumping, purifying, and distributing water and far too low to cover the long-term costs of maintaining water systems and making capital investments. Overall cost recovery in 2011 was estimated at 66 percent of the outlay. Two key issues affecting the shortfall—changes to tariff structures and to workforce numbers in the water sector—were seen as politically difficult to manage before the 2011 revolution, though the situation changed after the coup d'état in 2013.

Government actors and associated experts often characterize the public's resistance to paying for water as rooted in the popular conception of water as a gift from God and in outdated notions about the state's obligation to provide basic services to its citizens. State attempts to argue against perceived entitlements, whether of divine or historical origin, by using the logic of infrastructural legibility are bids to render technical (Li 2007) what are deeply political and morally inflected questions. State agencies seek to substitute one discourse for another when they urge residents to frame their increasing water bills as paying not for the water itself but for the infrastructure—the pipes and pumps that deliver water to doorsteps. This deliberate attempt to shift terrains from the object moving within the system to the system itself hinges on a project of cultivating the visibility of some things while reducing attention to others—a kind of background-foreground reversal (Hetherington 2019). Please pay attention differently, the state's argument goes.

However, my research in Ezbet Khairallah demonstrates that residents were deeply aware of the material and bureaucratic realities of water systems—a fact that should be clear after the discussion in chapter 3 of the installation of the city sewage system in Ezba. For years, residents lived every day managing the inconveniences of trenches, leaks, overflows, and expensive but weak pumps. Residents knew these systems intimately, and they saw them as dysfunctional. To be clear, inhabitants of Ezbet Khairallah did say that water is a gift of the divine, and they did believe that the state was required to provide water to citizens. But this was not the source of their unhappiness about the idea of paying some money for water services. Their discontent came from the real politics of the delivery and billing systems and the quality of the potable water they were paying for.

Work on the anthropology of infrastructure has addressed how infrastructure, and particularly the legibility of it, articulates specific political relationships. Early theorizations saw infrastructure as visible only in its absence or breakdown, reflecting the way that built environments and material or administrative systems can fade into the background of daily life (Star 1999). As Brian Larkin (2013) eloquently shows, however, the visibility of infrastructural systems spans the gamut from background to spectacle. Larkin suggests that the built environment is the material instantiation of ideological worlds, the put-in-place reflection of particular desires, frameworks, and agendas. It provides a foundation for the material and social possibilities of residents and a framework to which political and personal aspirations and affect can attach. This dynamic is visible in Timothy Mitchell's (2002) work on the Aswan High Dam in Egypt, which shows how mega infrastructure projects operate as potent political signifiers of postcolonial states' arrival on the world stage.

Both infrastructure and the citizenship that it purports to substantiate are fraught with contradictions, embedded surveillance, inconsistencies, reversals, and ambivalence (Anand 2017; Barnes 2014; Nucho 2016; Björkman 2015; Mitchell 2014; Larkin 2013; Von Schnitzler 2008). Residents of Ezbet Khairallah were deeply invested in managing their legibility to the utility system as an arm of the state; for them, visibility to the system both promised and menaced. Given the ongoing history of attempts to evict them, residents worried that the systems being put in place were not actually intended for their long-term use and were instead designed to serve future residents of luxury housing, which they feared would replace their neighborhood. The implementation of the sewage network enabled a closer accounting of who claimed which pieces of land, how big homes were, who lived in them, and what their history of connection with utility services—including their history of payment—had been. As Ezbet Khairallah was folded into the utility embrace of the municipality—into the pastoral care of the state—residents increasingly worried about what the state would do with the wealth of newly acquired details about the informality of their perch and particularly their lack of legal land tenure.

However, the health and social costs of living without reliable potable water and wastewater removal systems were and remain high—high enough to outweigh the risk of increased visibility to the state. Residents actively petitioned the Governorate of Cairo (GOC) for the implementation of a state system and paid for connections to be installed—all the while looking warily at whom and what they had invited into their homes. To be well connected is to be able to choose how one will be incorporated into utility projects and social networks. But residents were not always able to choose how or whether to connect. People did their best, then, to make uncomfortable connections fit as well as possible. By hedging connection—producing and managing ambiguity—residents navigated the possibility that connection could bring dislocation.

The risk of connection being uncomfortable cuts in multiple directions, and the GOC has been correspondingly ambivalent about its increasingly tight connection to Ezba. The Egyptian state writ large engages in practices of hedging connection as it seeks to articulate an agenda that sounds acceptable to multiple audiences, including citizens, international funders, and capital markets. State hedging makes possible a tactical deflection of claims that various entities within the state (GOC, the Holding Company for Water and Wastewater [HCWW], CAPWO) would need to address if the area were fully "seen." Hedging thus enables the state to maintain the status quo: the unequal distribution of scarce resources earmarked for the public good. Ferguson and Gupta (2002) discuss the breakdown of the notion of the state as a coherent entity that vertically encompasses its territory in a web of administration and oversight (see also Stamatopoulou-Robbins 2019). In Cairo representatives of various institutions selectively use hedging as a practice to enable their work as individuals or the work of the institution itself to proceed. I spoke with Dr. Magdy Allam, the representative to the People's Assembly for the Zahraa area, into which sections of Ezba fall. My meeting with him helped to clarify how hedging—in the form of selective misrecognition—characterizes the relationship between Ezba and many entities of the Egyptian state system.

Allam knew that most residents of Ezba did not have land tenure and had unofficial utility connections. Given my earlier experiences with officials dealing with the neighborhood (they were surprised that anyone was willing to visit there), I asked if he had campaigned in Ezbet Khairallah. He replied, "Many times, because I was elected local counsel for the Cairo governorate for twelve years, and I represented this area. I knew the area when there were only 10 houses, and I know the area now there are 10,000 houses." I asked how the houses in Ezba got started, and Allam said that gangs came in, divided the land among themselves, and then sold plots to those newly arriving to the city from areas in Upper Egypt.

"Who would do that?" I wondered. Allam replied, "Many people. People here use the resources of the country as if the government is not seeing them or sees them blindly. So, they do anything. The illegal actions are a lot. Every year you have a lot of houses being built, and no one asks them why. This is an old bureaucratic attitude, coming from Turkish and British colonization. They keep [this attitude and are] lazy in the government. This is a good rule. To be active in the government, you will be fired so quickly." Allam's description highlights the bureaucratic project of refusing to see, a blindness that has enabled both his own success and the ability of the state entities to manage the mismatch of the sizable demand for services with the limited or inconsistent resources for expansion. As the project of sowing connection, which I outlined above, shows, the density of bodies and the coalescence of local political will interrupted this blindness by forcing Ezba into visibility and contributed to the incorporation of Ezba into the municipal infrastructure. For the GOC, CAPWO, and the HCWW, the greater utility intimacy has provided more knowledge and more potential for social control in the informal settlements (Anand 2017; Von Schnitzler 2013), potentially enabling a long-term project to make this space available to accumulation by dispossession (Wahby 2018; Harvey 2010). However, this formal connection also has set a precedent, inviting residents to make more and more claims about the state's responsibilities to them.

In response to the discomforts of connection, the state has tried to redefine unregistered water consumption in Ezbet Khairallah from "informal use" to "theft." By shifting the frame, the state's formalization of connection to informal areas does not imply increased government responsibility. Instead, it deflects the blame for access, quality, and cost problems away from the GOC and CAPWO inadequacies or incompetence and places it instead on users' criminality. The state gives the appearance of addressing the water issues of the newly legible location but need not fully acknowledge the substance of residents' complaints. The Egyptian state has framed the problem as one in which citizens fail to properly "see" the infrastructure, pushing fundamental problems that residents have with pipes and bills to the background. The argument posits solutions—user education and state enforcement—that sidestep hard questions from residents about how and why these systems fail to serve them. By shifting the moral terrain from the government's obligation to the community's social responsibility, the GOC implicitly has authorized the use of force, moving from the slow violence of neglect to the infrastructural violence of water being turned off to the threat of razing noncompliant neighborhoods to the ground.

Water Is Costly

In a televised speech in February 2016, Egyptian president Abdel Fattah el-Sisi described the cost of producing drinking water and announced to the nation and the assembled governmental representatives that Egypt could no longer offer state subsidies for potable water services.

> To provide people with one [square] meter of water, how much does it cost? The cost to the nation is 160 piasters. So, we could say, it's 1.50 [LE]. That means 40 million LE a day. I'm telling the people who are hearing me now, for us to bring this water to arrive to you is costing Egypt 40 million pounds a day. Forty million just to make . . . drinking water. To treat it, how much will it cost? And we didn't talk about the station. How much did the station cost that we made to treat the water? And how will we guarantee that this station will continue working in a proper way? Why should we be aware of this? This talk is addressed to everyone: to the people who are in charge; to the Egyptian people who are

> receiving the service. We want you to know, you must know, and this information must always be present in your minds: when I am opening the water tap, I must know, the water that is coming to me, how much is it costing? So that conserving water will be something normal, and so that I don't say that the water is expensive for me, that electricity is expensive for me. (https://www.youtube.com/watch?v=JQcTHA_1B9M)

El-Sisi's declaration was a step in a decades-long process to reform the Egyptian water sector as part of a larger project of reducing government subsidies across a range of sectors. This move reflected the long-standing consensus among the government, international actors, and Egyptian citizens that the Egyptian water sector has worked poorly. El-Sisi's speech attempted to deflect potential popular resistance to increasing costs by focusing on infrastructural legibility, reinforcing the *produced* nature of potable water (Barnes 2014) to convince Egyptian citizens that water services must be paid for.

As el-Sisi hinted, people in Cairo resist paying for water. They have evaded or pushed back on payment for water by illegally connecting to state systems, failing to pay bills, and choking phone lines and customer relations centers with complaints about water service charges. The Egyptian Water Regulatory Agency estimated that in 2006, the average bill collection rate for the country was 46 percent, and that rate has remained relatively stable since. Local water systems throughout Egypt have not pulled in enough money to pay for their operating costs, and the Ministry of Finance has been forced to subsidize them.

Egypt had adopted a national water pricing plan in 1985 with graduated tariff increases that set a surcharge for wastewater at 10 percent of the water bill. However, those planned tariff increases did not go into effect, except in the case of the Alexandria water company directed by Nadia Abdou, which is its own story of exceptionalism. By 2004, the other fourteen regional water and wastewater companies, unable to meet their operating costs, had accumulated a deficit of about EGP 7.6 billion, with a debt of EGP 7.3 billion. These negative balances were blamed both on low tariffs and on inefficient operations. Overstaffing was seen as a particular problem, since the fourteen companies had a total of 74,000 employees.[3] In response to the increasing difficulty of covering these gaps, in 2004 Hosni Mubarak issued two presidential decrees. The first (135) established the Holding Company for Water and Wastewater (HCWW), which incorporated the fourteen water companies as subsidiaries and called for the establishment of additional governorate-level companies.[4] The second decree (136) established the

Egyptian Water Regulatory Agency, which came into being in 2006. These government agencies and their associated experts subscribe to the widespread international view of water as an economic good (Ballestero 2019b; Muehlebach 2019), which is frequently referred to as the Dublin Principles. These actors sought to apply the logic of aggregate costs and price pyramids to water tariff structures to better reflect the cost of production in an attempt to retrofit water services so as to bring citizens into a proper relationship with the government.

According to experts, however, the Dublin Principles ideal is hard to achieve in Egypt, given the cultural value of water and given that people have particular expectations about the right to work (making it difficult to reduce overstaffing in water treatment plants). The twenty-first century has seen a series of national and bilateral international aid development projects focused on reforming Egypt's water sector. I interviewed David Osgood, the country manager for CH2MHILL (an engineering subcontractor of Chemonics International working under USAID), in which he explained that the water tariffs in Cairo were some of the lowest in the world. According to him, the financial difficulties of the sector were

> just political. The government just doesn't have the political will to increase tariffs because there is a sense here that water is a gift from God, and you shouldn't charge them for it. Plus, there is still sort of remnants here of socialism from the days of Gamal Abdel Nasser, and so people feel the government should take care of this, and [they] shouldn't have to pay for it. So, you know, [the government] fight[s] those fixed ideas in people's minds. . . . They pay a lot of attention to what they think the people think, okay? Because there is always the threat of riots here. It doesn't happen very often, but every once in a while, it happens. They raised the price of bread twenty years ago, and it really set off a big problem. They are conscious of those kinds of things.

By "those kinds of things," Osgood meant resistance by the population in great enough numbers to potentially destabilize the political and economic status quo. Using Chatterjee's (2004) terms, the potential for extrajudicial violence is part of the political society in Egypt. This is not ballot-box democracy, a hollowed-out import at any rate, but concern about violence that could undercut the legitimacy of the ruling regime. Karen Piper (2014) argues that thirst and the long struggle for access to water services were important factors in the 2011 revolution in Egypt.

Residents in informal areas had long been fighting a municipal water system that prioritized service to expansive, high-priced new buildings on the desert outskirts of the city at the expense of existing urban spaces. But during my interview with Osgood, the threat of violence seemed remote. Less than a year after that conversation, however, "those kinds of things" took over Tahrir Square for the first of what would be several reversals in the following decade. In the early 2020s, violence again seemed remote; public anger had dissipated, narrowed, and been submerged, for the el-Sisi regime had the experience of riding the wave of the political situation into power and had worked to limit the possibility for mass mobilization (Flahive 2018). However, the risk represented by popular discontent is never entirely absent, and as el-Sisi's 2016 speech indicated, his regime has actively worked to reduce public discontent around water services by building on a long-term project to ask that citizens focus on the produced nature of water. This was a project of hedging through shifting the framework within which people were reading their payment for utility services.

During a morning visit to the Ain el-Sira complaint center in late 2010, an official from the Cairo Water Company asked me to participate in this hedging project by acting as an intermediary between the water company and the residents of Ezba. He wanted me to assist with the project of reframing by explaining to residents that the government was charging not for the water but for the cost of getting it to their houses:

> You know them [people in Ezba], you speak to them. Please explain this to them! Explain to them that we are not charging them for water. We are charging them for the cost of bringing it from the river Nile to their homes. Is it free to pump the water? Is the pump free? Is it free to purify the water? To build the pipes and pay the people to bring the water all the way to their faucet? No. It costs, and they must pay their share. People in Egypt don't pay much for water, not like the rest of the world. In other countries it costs a lot, and here it's almost free—like we are giving it away.

In this moment, the official was attempting to turn me into a conduit of framing. Please help us train the residents to pay attention differently, he said—to see pipes rather than water, to see informal use as theft. This man worked in a complaint center for the water utility. All day, every day he heard complaints precisely about pipes and prices. His attempt at reframing sought to mitigate the discomfort of the government's utility connection with Ezba—to hedge the connection, to make it less intimate, less binding, less fraught with expectation. The complaint centers themselves were created as part of the larger project of reforming the water sector. They were ostensibly initiated to mitigate public displeasure, but perhaps more important, they served to cultivate new kinds of responsibility for the water system among its users: to report leaks, to help to make water delivery more "efficient," and to participate in the functioning of the utility systems. This man sat between two different framings—informal use versus theft—and his encounter with me must have seemed like a serendipitous opportunity to reduce the power of the Ezba citizens' frame and hedge an uncomfortable intimacy.

I had a conversation with a USAID staff member later in 2010, and he repeatedly compared people's willingness to pay for communication sector services to their unwillingness to pay for water. Like the water sector, the communication sector had also been newly regulated and restructured to consolidate operations. However, services had dramatically improved in the communication sector in a way that they had not in the water sector. According to this USAID official, this was because people were *willing* to pay for communication services. Inhabitants in Ezba, he told me, would willingly pay EGP 100 a month (at that time, roughly equivalent to USD 20) for their phone and use it for pleasure—chatting with relatives, buying ringtones, sending money to friends. But paying for water was a harder sell.

This comparison between sectors was misleading, however. It is doubtful that residents of Ezba saw a direct trade-off between phones and water, and this official's categorization of water service as essential and communication service as nonessential missed how residents actually used these services, how well the services functioned, and who they believed should be responsible for providing them. The time and money spent connecting with others is neither entirely frivolous nor entirely instrumental but is instead fundamental to connected lives. It is part of maintaining networks of love, reciprocity, and possibility. These connections are the phatic labor that Julia Elyachar (2010) identifies as a whole infrastructure of linking that makes other things "work." These connections offer the hope of help when an emergency surgery looms, or the chance at a job in a distant city passed along by a friend of a friend.

People recognize that when there is no money, there is no cell phone credit, and landlines—where they exist—will be shut off for nonpayment. No one will die without their phone. But how is one to live without water when there is no money to pay? Residents of Ezba saw the provision of water as a responsibility that the state has toward its citizens. If they were recognized as full citizens, they felt, they should also be guaranteed access to water and at reasonable and steady rates.

Paying for Water

As I detailed in chapter 2, acquiring water in Ezbet Khairallah has long been a source of consternation. The physical toll of carrying water by hand and head was paid by women in the area. Tanker truck provision duplicated one of the ubiquitous inequalities of water access globally by making the poorest—such as those in informal areas without access to city services—pay the highest rate for water (Ballestero 2019a).

Hajja Basma had lived in Ezba for twenty-six years, and her daughter Fatma had lived the entirety of her twenty-two years there, so they had seen it all. Describing the situation that she'd lived through, Hajja Basma said, "At first there was no electricity or water or sewage. People carried water from nearby. [Piped] water came to Ezba about ten or fifteen years ago. Before, it used to be that a truck came around and sold you drinking water for about a quarter of a pound for a jerrican [roughly ten liters] of water, or a couple of pounds for a big barrel of water for building the house." She and her daughter stressed that the barrel water was not for drinking; it was used only for construction or for household chores and bathing. According to Hajja Basma, Ezba finally got water thanks to Fayda Kamel, who ran for Parliament. When she won, she brought water to the area: "Fayda Kamel and some other man, whose name I don't remember. They stood together for election and when they won, they brought water in, created a water company. But you can find all of this history on the internet," she assured me. "It will all be written there." Hajja Basma remembered the shock after her meter was put in and she was told that she owed something like EGP 3,000. She said, "I couldn't pay it all at once, but little by little, gathering money from my sons, I was able to pay it off and finally I got a meter, and now [I] only pay for what I use. A meter like they have for electricity. Now each apartment [in the family building] has its own meter."

As the story of Kamel shows, utility services in Ezba were often obtained through votes-for-services swaps. Community members petitioned the state through demonstrations and public gatherings to get services, and in response political candidates promised to implement first electricity and then water utility services if residents voted for them in parliamentary elections. As Gandy (2004) points out concerning the long arc of water infrastructure in Mumbai, those with the greatest need and the most limited access to centers of power are often the most vulnerable to the political machinations of ballot-box electioneering. In my conversations with officials, it became clear that the political activism of the people brought results: pressure from residents in places like Ezba grew intense when the number of people and the unofficial drains on water networks reached such levels that they could no longer be ignored. When I asked how Ezba got official permission for government water, one employee of the Holding Company for Water and Wastewater said, "The government had to respond to what they found in front of them!" Residents of the area had made themselves conspicuous.

In exchange for implementing official wastewater connections in Ezba, the Cairo governorate insisted that water meters be widely installed alongside. Until this time, many homes in the area did not have water meters and either avoided paying for water entirely or paid what someone referred to as the "*garas* bill"—a small bribe paid to the collector who rang your bell (*garas*) to not report you. The government's push to increase the number of water meters introduced anxieties. What exactly was being measured? Who was keeping track? Who was owed the money? How did the measurement move? One woman who had not yet installed a water meter asked me if the government would be able to tell how much water she was consuming by measuring the sewage output from her house. She was afraid that her unregistered water connection would be identifiable and measurable that way. Because what goes in must come out, she worried that wastewater might become a telltale sign of "theft."

A longtime resident of Ezba, Um Sara had weightier concerns. As the government continued to push the narrative of theft, Um Sara turned over in her mind the religious implications of this secular reframing. When we discussed the installation of water meters, Um Sara was struck by a thought: "If we are paying for water now, and they say we have been stealing it this whole time, can you imagine the [moral] cost? Think of all of the millions of drops of water over the last fifteen years—to be held responsible for all of them?!" The moral burden of having "stolen" what amounted to hundreds of millions of drops of water over the years was overwhelming. How was that water to be accounted for? Was each drop to be counted as an individual deed, a debt of bad behavior that must be made up for by good deeds before meeting one's final divine judgment? The state's redefinition of the use of water had the possibility of reordering Um Sara's perception of her actions. *Use* became *theft*—an action carried out against Egypt's interests (the state's framing) and creating a moral debt (the religious framing). The cosmological consequences of these questions hung heavily. Under the new discourse of water "theft," simply having existed in Ezba could damn you. The new discussion of theft also came into tension with older ideas among residents of Ezba who had never fully incorporated the moral discourse that the state was attempting to center: a shift from state responsibility to social obligation by residents, from water as a gift from the divine to payment for the network bringing it to your doorstep.

While the state sought to reframe informal water use as theft, residents of Ezba saw the arbitrariness of water bills as a challenge to the legitimacy of the state's claims. On the surface, it appears that the tariff structure favors the interests of Egyptian consumers. Residents pay a graduated cost per cubic meter of water based on usage amounts and the category of the user (residential, commercial, industrial). As is often the case, the lowest one or two tranches of the structure is designed to protect very poor people. When compared to other global water prices, Egypt's tariffs were quite low. In 2009, the bottom tranche cost EGP 0.25 per cubic meter and the top tranche EGP 0.35. Yet this structure did not always translate into affordable water for residents, who contended with bills that did not seem to be correctly calculated. When I asked David Osgood how the tariffs were calculated, he replied:

> It's what people are expected to pay, and at this point it's not really based on anything except somebody's judgment of what people are willing to pay. . . . The sacred part of the tariff schedule is the domestic. . . . So for sure, that lowest category is absolutely sacred. Nobody is going to be able to raise that. In most cases, the other two classifications are untouchable also, but there have been some situations where those have been raised in a kind of stealth manner. They never announce it. People just get a bill, and they say, why is my bill so much higher this month than it was last month?

Osgood was laying out a state version of hedging. Rates for some in the higher tranches of payment were raised without notice or acknowledgment. However, as residents of Ezbet Khairallah told me, those on the higher rungs of the payment ladder were not the only ones to face surprise costs. In their everyday engagements with the state around water issues, from the delivery of the bills to the complaint centers, residents of Ezba did not see themselves treated as part of the protected class of users—those in the "sacred" rate categories. Their water costs did increase—and unpredictably. They therefore experienced attempts to raise water costs as manifestations of the government's disinterest in their well-being, at the least, or possibly as corruption or incompetence. As with Ali's electrical bill (see chapter 2), the water bills that Ezba residents received often fluctuated without any discernible reason; accounting mechanisms did not seem to increase accuracy but simply added to costs. Because the bills appeared arbitrary, residents rejected the rationale that they were necessary for an improved system, and they evaded or circumvented the systems that they believed to be predatory and indifferent to their welfare.

Residents saw the water bills as arbitrary for two reasons. The first was that the bill totals did not match people's usage. Bills either showed wild swings in cost from month to month, despite no significant differences in a resident's actual usage, or arrived without a meter reader ever having visited to ascertain genuine usage. Um Amina told me that the first bill her family received after installing a water meter charged them EGP 400 without a meter reader ever visiting their home. Over the course of the first half year of being metered, the bills amounted to EGP 3,000. She and her husband went to the complaint center to ask for an investigation into the high cost of their bill and were told that the center would only investigate after the family had paid the complete amount due. They considered joining an informal saving association in order to pay off the arrears but worried about being able to pay the bill before the water was shut off. When I asked if they could take out the meter and reconnect informally to the system, Um Amina responded that now that they had been officially connected, the state knew that they were there and would know if the meter were removed. They had become irrevocably legible to the state. The potential ramifications of any "theft" of water were serious: there were discussions in Ezba about possibly being jailed for tampering with water meters or for connecting to state systems without a meter.

The ability to enforce compliance in Ezba will likely only increase, for the Egyptian Water Regulatory Agency has asked the Egyptian judiciary to empower workers to bring legal action against individuals for water usage violations. Um Amina's husband told me, "We put in the meter because we wanted to do the right thing, but it ended up costing us dearly."

The second reason that residents saw the bills as arbitrary was that the price per cubic meter was determined either without explanation or incorrectly. Many low-income Cairenes live in multistory buildings with only one meter per building, and the water rate increases with the volume of water used. A small store on the ground floor of a residential building potentially flags that meter for commercial rates; this means that low-income residents living in a large apartment building can end up paying the higher rates meant for larger-volume consumers. This also affects sewage prices, which are charged at 10 percent of the total cost of the water for that month, so the effects of billing errors compound. In 2016, government officials announced a 25 percent increase in water tariffs. By *Al Ahram*'s calculations (Howeidy 2016), this 25 percent increase could in fact add up to a 158 percent increase in users' costs.

Meters for water were also suspect as part of the "recognition" matrix, which became more necessary (and more unavoidable) as government services expanded into the area. Recognition, like connection, is a double-edged sword: it holds out the possibility of legitimating land tenure claims, but it also creates unwieldy costs for low-income residents, opens them to surveillance, and possibly sets the stage for dispossession. Many households in Ezba had not installed meters when the water pipes were installed fifteen years before the wastewater system, though a few had. Now the cost of meter installation had risen dramatically, pricing some of the most vulnerable residents out of the market. What had cost EGP 700–1,000 when Teta installed her meter roughly five years previously cost EGP 3,000–6,000 during my fieldwork in 2009–2011. As the implementation of the sewage system increased demand for and the visibility of potable water meter installation, the price for the meters had risen. Despite the higher cost, residents felt that they now needed to be on meters, because the formalization of water and wastewater services by the GOC was touted as a key component in the move toward granting land tenure to the people living in Ezba. Residents would be able to provide a history of utility bills as evidence of their land claims through the court case described in chapter 2. The expansion of the GOC services thus offered both a carrot and a stick: the opportunity for recognition by state entities but at increasingly prohibitive costs and with uncertain outcomes.

As the possibility of land tenure came into view, many families were being left behind. They had not been able to install expensive meters and pay for the high cost of collectivized bills for either water or electricity. The dramatic fluctuations in costs for both the water itself and the associated equipment called into question the broad narrative of progress—the dream future of a safe and convenient Ezba where strong nets of infrastructure contributed to a pleasant place to live. For some, this dream future darkened and became dystopic: they saw their more prosperous neighbors walking forward into that future while they simply fell further behind. Residents in Ezba were concerned that the promise of infrastructure was *ṣābūn*—a lie to grease the passage toward greater state knowledge about their land, family, and home. This could mean the wholesale removal of their neighborhood; it could also mean smaller-scale tragedies of residents unable to keep up with the new improvements, whose carefully sown connections could become uprooted by gentrification. This difference between neighbors in the spectrum of utility connection was a question that was open to speculation by other inhabitants, though some wished to keep such information private, as the next story illustrates.

Selling Soap

The area in front of Um Amina's soap shop was a place where people often gathered to discuss issues big and small. The shop was a converted room on the ground level of her two-story family building, a brick and concrete structure wedged cheek by jowl with others of similar construction along a narrow side street. Most of the items sold were intended for quick consumption, and the margin of profit was small. Um Amina was able to earn a moderate income from this store for two main reasons: she didn't have to pay rent, and her husband owned an auto body shop. She therefore had access to cheap transportation to bulk markets in the center of Cairo, where she went every few months to resupply.

The household cleaning products that she offered were a necessary component of women's work—caring for home and family—in the context of the material situation of Ezbet Khairallah. The lack of clean, dependable water had many ramifications for health and household cleanliness, multiplying the care work to be done. Um Amina's soap shop was one link in the larger chain of managing the labor of scrubbing rugs to clean off "dirty dust," washing laundry that often had a ring of sewage-tinged mud at the bottom hem, and cleaning food, dishes, and children as well as possible. The shop was also a personal way for her own family to make things work. She earned enough money for school fees, to purchase items for her daughters' trousseaus, and to cover household costs when the income from her husband's workshop dipped, as it did from time to time. Um Amina spoke with me about the obligations and effort required to inhabit her world as well as possible (Tronto 1998).

As Clara Han (2012) points out in her discussion of how moral projects unfold in everyday life, care is a problem rather than a given, something that is worked out in the context of multiple needs and limited resources. Han calls attention to a practice reminiscent of hedging, noting how her interlocutors used the conventions of testimonial (*testimonio*) to sometimes make themselves and their needs present to others and at other times to make their needs opaque, allowing room for others to get by. In hedging, communication is designed to mediate the tension between social expectations of mutuality and material realities of scarcity, but in hedging practices by Ezba residents, the goal was specifically to produce ambiguity in order to manage social relations often built through phatic labor.

When I arrived for a visit at Um Amina's store one day, I found a group of four ladies from the neighborhood standing out front. They were discussing circulating rumors that the cost for potable water meter installation was increasing from EGP 1,000 to EGP 6,000, seemingly overnight, and they were speculating that the change in price was due to the ongoing implementation of the wastewater system. The eldest of the assembled women, Um Maryam, said that everyone had to pay for the new sewage system in one way or another, implying that the Cairo Water Company was taking the opportunity offered when putting in one system (sanitary drainage) to increase prices for components of the other (potable water). Um Fatma chimed in, saying that her own potable water meter had been installed for a very long time. At that point, Um Amina challenged her, insisting that Um Fatma did not have any such meter.

"No," Um Fatma replied, "we installed it when your mother-in-law was still living in the room behind you," gesturing into the house behind the soap shop. All of the other women present simply left the statement to hang on the air for a moment. It was clear that no one wanted to challenge the factuality of Um Fatma's assertion, but there was no general agreement that she was making a truthful statement. Finally, someone changed the subject by offering the name of another neighbor who had recently installed a potable water meter. The ladies visiting the store then collaborated to create an impromptu list of names of those around them who they thought had installed meters.

After they left, I asked Um Amina if she had a meter, and she said no, she had never installed one and that very few people in Ezba had meters. "Did Um Fatma have one?" I asked. Um Amina shrugged. Whether or not Um Fatma had a water meter was unclear, which was precisely the point. In this case, she was hedging—giving the assembled ladies the *ṣābūna*.

When Ezba residents talk about giving someone the *ṣābūna* (literally, soap), they mean a facile answer or small lie that might deflect the person, send them in a different direction, in order to avoid distributing information that they don't wish to share. In claiming that she had had a meter installed, Um Fatma was working to manage her legibility—others' knowledge of her circumstances—by giving the *ṣābūna*. Who knows what it would cost her for these women in her network—or a newly arrived anthropologist—to know about her meter status? This is an example of how and why people parse and hedge connection. When is it useful or dangerous to tell the truth of one's situation? What are the stakes? Um Fatma was faced with competing, ambiguous risks. If she didn't have the meter for a long time, she might be open to official consequences for water "theft" or could undermine projects to use utility connections to substantiate land tenure claims. Conversely, if she had a meter already installed, did that mean that she had disposable income at a time when others were struggling to afford to install their own meters? Um Fatma made a calculated assessment in the moment about the likelihood of various outcomes, with the understanding that future interactions might call for different assertions and that the rules of the game might change as governmental laws and uneven enforcement shifted the hazards. As this story shows, the project of hedging is about not only managing connections with the state; hedging mediates connections between residents as well. Some part of Um Fatma's concern might have been about not wanting to get reported to the state, but some was also about fending off possible future problems with her social connections. In conditions of uncertainty, even seemingly innocuous things might be the subject of hedging because the directions of threat in the future are unknowable in the present, and perhaps because some forms of attention may be unsettling.

During my time in Ezbet Khairallah, I asked a few close connections if they would be willing to share their water bills with me for my research. No one ever said no, but no one ever produced one either.[5] I was given vague promises to share them at a later date, a delaying tactic (*yamaththul*). I saw several bills on tables or near a TV set in people's homes, so they were not generally hidden as objects. But to have someone, even in the context of research, take a picture of a bill with details like name, address, and amount owed would have been too definitive. It would have bared too much, been too uncomfortable. This was not just about the public circulation of private information, but about the preservation of some control over what can be said about one's water meter and bill. As with Um Fatma's meter, when information is not explicitly shared, there remains some wiggle room—a freedom to be flexible about the truth. When others gain too much *certain* knowledge, it limits the possibility for maneuvering. I read residents' hedging on sharing copies of their bills as not just about revealing their personal information in local or international contexts, but also about the stakes of maintaining the possibility for hedging as a broader practice of managing uncertainty.

Hedging is part of the social infrastructure that undergirds and maintains the worlds that residents co-create. Hedging is a practice of maintaining social channels, protecting relationships while leaving room to avoid others' attempts to activate instrumental flows through them. In *For the City Yet to Come* (2004a), AbdouMaliq Simone shows that webs of social relations are a kind of social infrastructure that supports people but at the same time can drag them down. These networks are a drain when there are too many demands and too little of value to disperse among the lines of connection. One quality that Simone recognizes in these webs under duress is a perceived need by many to play everything very close to the vest to protect fragile possibilities to mediate against uncertain futures.

An example of this in Ezbet Khairallah was an elusive NGO that I was often questioned about during my first months in the area. In one instance, that NGO was responsible for putting in a water pump and connecting a woman's house to the water pipes. As a widow, she was understood to have very little financial capacity, so the installation of expensive infrastructure at her home was seen as curious. People finally went up to a worker during the installation and asked them how she was financing all of this work, and the worker told them it was a project sponsored by an American aid organization. Her neighbors bitterly resented that she had never told them of the NGO's existence or her affiliation with it so that they might also benefit. The widow was playing *istiʿbāt*, and her neighbors were trying to parse her hedging project. These neighbors told me, "We don't need anything. We are very lucky. We have a house and food and are fine. We just wanted to know if you had heard of or had any information on this organization." They were trying to locate some previously veiled possibility through me as a new possible node of connection. I might have been a conduit to the "American NGO" as an American researcher. The careful management of knowledge, or the appearance of knowing something, was again at work.

There is constant speculation about who knows what, when they learned about it, and what such knowledge is worth. When one has valuable knowledge, selective evasion as a form of hedging is one way of making sure that such knowledge remains useful.

Conclusion

Being well connected in Cairo requires building and maintaining strong social networks with neighbors and advocating for state services, but these requirements also open Ezba residents to possibilities of violence, extraction, and dislocation. As I have demonstrated, residents attempt as best they can to function as Juha—meeting situations with facile skill and humor in order to hold open possibilities against the many pressures that threaten to collapse any room to maneuver, or even room to breathe.

Dr. Allam's story articulates the political situation in which Ezba residents found themselves as they sought to gain services while maintaining some kind of balance in the potential costs of gaining recognition. In the face of deliberate evasions, misrecognitions, and disavowals by state agencies, people may choose to force themselves into visibility (see Wahby 2021). Indeed, at one level, it was through the sheer force of massed bodies, bricks, and piles of concrete—and the potential for them to overflow, exceeding the limits of the space and even rebelling—that Ezba residents were able to get piecemeal accommodations from various government departments.

Officials have made it clear that they are pressured by the presence of places like Ezba when the volume of people and the unofficial drains on water networks reach such levels that they can no longer be ignored. In response to this pressure, the state hedges. It reads individual water meters, but sends out collectivized bills; it measures leakage (Anand 2017) to account for the water being drained through unregistered connections, but never addresses why informal areas are tapping the water; it charges EGP 15 for a water bill one month and EGP 200 the next month without explanation. All of this enables state actors to deflect potential critique, manage dissent, and continue operating on the logics of appropriation.

On the other hand, hedging provides marginalized urban residents with a set of discursive and practical tools to deal provisionally with state actors that can seem quite inflexible to those experiencing government's effects. Through hedging, residents of Ezbet Khairallah deflected and managed the ever more intimate control of private spaces that sewage pipes and water meters entailed. They practiced hedging among neighbors to create and maintain channels created by phatic labor (Elyachar 2010) through limiting the things that might move through them. Hedging gives wiggle room within the moral and material problematic of care in pernicious surroundings. The play in the system allows people to get by each other, to carefully manage the appearance of knowing things, and to restrict the visibility to others of limited resources.

Chapter Five

Sensing Connection

It was evening when I got off at the Dar el-Salam metro station in 2009 with a young male friend from out of town. Walking in the direction of the main road to find a microbus, my companion stepped in a puddle of water. He stopped cold, raised his foot, shook his sandal sharply, and then headed in the direction of the nearest pharmacy. "We need some rubbing alcohol," he said. I replied, "I have a tissue. Would that work?"

"No, I need to clean it with alcohol. Who knows what's in that puddle? Maybe swine flu."

At the height of swine flu, this throwaway comment indexed a larger set of anxieties about the dangers posed by unknown water and the permeability of bodies to infection. It also condensed worries that were specific to the kind of mixed-religion area that Ezbet Khairallah is, where Muslims live alongside working-class Christians, who might be keeping pigs—or, more broadly, might be susceptible to swine flu contagion, which was understood to run along denominational lines.[1]

Inside the pharmacy, my companion cleaned his foot and sandal with disinfectant and then asked for the easiest walking directions to Ezba. Surprised, the pharmacist asked, "Why would you want to go there?"

> TESSA FARMER: I've heard a lot about the water issues in that area and wanted to see it.

> PHARMACIST: You don't want to go to Ezba. It's filthy. There is garbage piled up higher than your head. The people are all covered in filth from head to toe.

> TF: Have you ever been to Ezba?

> PHARMACIST: No, why would I go there? They are all thieves and drug addicts and prostitutes. People there have scars across their cheeks [gesturing with his thumb from ear to mouth] from the knife fights. The streets are covered in sewage. Ezba is not a place to visit.

In this story of a puddle, my friend—an outsider to Ezbet Khairallah—used his own material senses, his own body as an instrument for engaging with the water in that place through the paradigm that he had. The narrative of Ezba went like this: it was a place of danger, of abandonment, where the streets were paved with sewage. There, encounters with water threatened both material ills (the sort that come with contact with human waste) and the more figurative ills that come from the affective response to matter out of place (Douglas 1966) in a dangerous locale. My friend had no connection to the context of the water in a puddle near Ezbet Khairallah, and this heightened his perception of danger and changed how he judged and interpreted the encounter. He knew that water quality matters, but as an outsider to Ezbet Khairallah, he couldn't assess the safety or danger of water with the same nuances as a resident.[2] This exemplifies the almost universal projection that water in certain "elsewheres" is inherently dangerous—a notion captured in the classic traveler's aphorism "don't drink the water."[3]

The pharmacist and my friend used their bodies as instruments to both assess and map the quality of water—and, by extension, the quality of life—in Ezbet Khairallah; they used their senses to evaluate the "well" of being well connected. For residents, too, assessing the quality of water in Ezba is an embodied act. People extend their sensory processes to evaluate whether water is consumable, whether it poses a threat, and whether it has potential for different kinds of thriving. In Ezba, water exists along a spectrum of potability and approachability, from water that contains too much sewage to be consumable, to water that contains so much chlorine that it is potentially carcinogenic, to water resting in clay pitchers as a gift, to water pooling in street intersections. In Ezba, water quality is made legible by people who are intimately rooted in place. They are connected well enough to make water understandable in each particular location and context.

In technical terms, water "quality" is a description of the condition of the water in a specific locale. Factors such as chemical content, physical condition, and biological characteristics are evaluated to determine the suitability of some water for a particular purpose (e.g., drinking or swimming). Technical water-quality indicators, such as dissolved oxygen, pH, salinity, and nutrient (e.g., nitrogen and phosphorus) content, and measurements of toxicants, such as insecticides, herbicides, and metals, are by and large not available to residents of Ezba.[4] But they have their own processes of assessing water's acceptability for various uses: smell, sight, hearing, taste, and touch. Residents use these sensory processes to qualify water—to get to know the characteristics of a particular water source rather than assessing it simply in terms of the binary *good* or *bad*. The quality of water in Ezbet Khairallah exists on a spectrum rather than as a dichotomy of potable water versus grey or black water, or healthy water versus contaminated water. Residents seek to know if water has enough of particular traits (temperature, taste, smell) along a spectrum of acceptability for particular tasks.

The water in question is more than simply the water that people consume; it is also the water that residents step over in the streets, manage at their doorsteps, and watch sluicing over fish at the market. Sensing connection in Ezbet Khairallah is a project of mediating the relationships that connect various forms of water and the relationships between people and their water sources. It uses the body's permeability and porosity along with the contextually trained bodily capacities of smell, taste, and physical pressure, among others. To learn to sense connection in Ezbet Khairallah, one must refine the sense of what will harm or heal, what tastes appealing, what might cause disgust. At a certain level, this is a universal project; the separation and management of dirty and clean water are global and age-old problems. A reader in the United States will sense the difference between water running in a sink and standing in a toilet bowl, but the produced "backgroundness" of water infrastructure (Star 1999) in the United States makes the project of water-quality discernment fade into the mundane.[5] In Ezbet Khairallah, water was not backgrounded; potable water was something to be constantly cultivated or mediated, as was the proper disposition of wastewater. Thriving in Ezba required the developed skill of sensing water.

Connecting with water in this way relies on sensory density, as people's trained senses interrelate to produce judgments. Residents use their noses to discern if their potable water has an admixture of black water or, conversely, too much chlorine. Their sense of smell also can tell them about the condition of the roads, for in the arid area of Ezba, mud is almost always caused by sewage overflow.[6] They use their eyes to assess the material reality around them, catching details of water's impact on streets, houses, and bodies. They also visually judge water in containers, checking for the appearance of dirt and contamination. Inhabitants use their ears to listen for the rhythm of flow and stoppage, to hear complaints from each other about the conditions they live in and the potential solutions for the problems that they face, or to measure leftover water in a defunct septic tank. They use their taste buds to assess potable water as it moves along a spectrum from life-sustaining to contagion vector. People experience the flavor of the water that they are familiar with and distinguish it from the taste of water sourced elsewhere. They evaluate the water in a container for freshness (how long it has been sitting), and they taste the things that have been added to water to sweeten or improve its flavor, such as mint, cut fruit, or rose water in *sabīls*. Residents use their sense of touch, their skin, to gauge the temperature of water and sense its potential to purify both morally and materially. Smell, sight, hearing, taste, and touch all overlap in the project of connecting to water.

Anthropological studies of the human senses have shown that bodily perception is fundamental to the formation of identity and subjectivity, to power relations, and to placemaking (Seremetakis 1994; Stoller 1989). Sensory perception in cities is an important object of attention, although urban design literature often studies sound, smell, and vibration only as "environmental stressors" (Henshaw 2014). Some scholars do attend to the importance of sense data in urban spaces. Alisa Perkins (2015), for example, shows that the call to prayer in the Michigan town of Hamtramck was an important aspect of producing a welcoming locality for Muslim immigrants from Yemen and Bangladesh, but some Polish Catholic residents experienced it as invasive, as "inescapable," as an inherently foreign sensory experience. Likewise, Hirschkind (2006) examines the prevalence in Cairo of cassettes of Qur'anic recitation and popular exegesis, showing how they shape the ethical landscape of the city through sound. Jamie Furniss (2017) documents his own bodily experience of fieldwork in Cairo among the Zabbaleens, which was shaped by the smell of garbage; an early talk on this research by Furniss sparked a larger project by Victoria Henshaw (2014) on designing city smell environments (see also Caton 2017). Poet Mahmoud Darwish (2011) indexes nostalgia by describing the olfactory bouquets that are characteristic of particular cities throughout the region, and he suggests that the absence of a distinctive scent profile makes a city unreliable, perhaps because that makes it less imprinted in one's memory. In the poem "What Is Lost," Darwish describes Cairo as smelling of ginger and mango. In this chapter I follow residents of Ezba as they leverage their entire bodies for the project of sensory placemaking, including but not limited to scent. I show that their trained senses are part of what makes residents well connected in Ezba. The overlap between city smellscapes and city waterscapes is an interesting question.

Stories of water are important to theories of identity (Strang 2004), can function as prisms for understanding theories of the social (Helmreich 2009), and are an important object of attention for science and technology studies. In an evocative 2010 examination of the sensing of water in an Omani town, Mandana Limbert demonstrates how changes in water distribution and management systems in Bahla intersected with sensory regimes tied to transformations in notions about women's bodily privacy, expectations of mutuality among neighbors, and nostalgia for the aural experience of water. Understanding water quality is a process of interactions between locally trained sensory experience and scientific knowledge.[7] The potability of water has a lot to do with the fact that water is hospitable to nonhuman life, and Ezba residents are sometimes *too* well connected to the organisms whose presence in their lives is made possible by water. As I show in this chapter, even what is tested for is mediated by local ideas about what is good or bad, what is likely to be present, and what the spectrum of water quality may look like.[8] The literature on the sensory experience of place and of water quality helps clarify what it means to be well connected in Ezbet Khairallah.

I contextualize the discussion of how Ezba residents sensed water with news reports of water-quality scandals in Egypt in the early twenty-first century, as well as an extended interview with an official at an international organization working on water in Egypt. I elucidate the many ways in which connection itself has quality; you can be connected poorly or connected well on the larger Egyptian stage. I next move to ethnographic vignettes of a failed attempt to get water testing done in Ezba, detailing how projects to know water are mediated both by interspecies relationships, such as between people and bacteria, and by larger cultural frameworks for making sense of contagion. I then center an interview with a younger resident of Ezbet Khairallah as she discusses the ways in which she learned to know water quality through sensory processes, detailing how her senses have been trained and informed by local historical knowledge of sensory landscapes.

Sensing connection in Ezba is about discerning a spectrum of water possibilities and assessing how well connected residents are more broadly to social networks and to the larger framework of municipal services in Cairo.

Connection as Quality

As of 2015, official statistics showed that 99 percent of the Egyptian population had access to improved potable water, which the World Health Organization describes as a source that adequately protects the water from outside contamination, particularly fecal matter.[9] As with the wastewater systems I examine in chapter 3, potable water availability and quality vary in urban and rural contexts, and the statistics often hide significant differences in access and quality along socioeconomic lines. In rural Egypt, where there are places with no formal water distribution networks, water is often sourced from unofficial, untested sources, which may be of poor quality. In these rural areas, residents often pull water from irrigation canals that have been contaminated by agricultural fertilizers, pesticides, and sewage, or from contaminated groundwater sources. In urban spaces like Ezba, water is often insufficient: it either does not go to an area at all, or it does not go regularly or in large enough volumes. Further, the material networks through which the water travels from water treatment stations to homes create many opportunities for the introduction of contaminants from cracked pipes, overflowing septic tanks, and poorly maintained water storage units on rooftops. The pumps that have been installed to increase pressure to particular homes may themselves crack pipes, and cuts in service often allow wastewater to infiltrate the system.

As previously discussed, there are many problems with the existing water systems. End users have low water pressure and limited supply; there is significant overstaffing; employees at regional water companies, water pumping and purification stations, and wastewater treatment stations are undertrained and paid low wages; and there are technical issues with balancing water chlorination. A lack of system maintenance often exacerbates supply and quality problems, as with Ezbet Khairallah's long-term issues with the contamination of potable water by black or grey water. There also are questions about the capability of those working to meet the standards for quality production.

One of the issues stemming from the lack of trained staff is the handling of chlorine in the water purification process. This is something that Chemonics has been working to address through a certification program. In most developed countries, treatment plant management requires certification and a set of core competencies. David Osgood said:

> There is almost one serious incident a year, in terms of water, where people get sick and die someplace. The last one was, I'm not exactly sure, but just north of

> Cairo. The year before it was in Dakahlia, in Mansoura, in an area outside of Mansoura, where the sewage somehow managed to contaminate water supplies. So, water quality continues to be an issue. You know, in the last project when we were working with the Holding Company and working directly with water or the treatment plants, we were doing assessment, there were some reasons to be concerned in some cases with the quality of the treatment. But probably overall it's okay, but you will find instances when it's not okay.

In addition to filtration for floating elements, chlorination is the primary method of treating drinking water for living organisms, and people in Ezba often complained about the balance of chemicals in the water. The complaints ranged from the relatively mild (the smell and taste of the water) to the severe (liver and kidney problems from prolonged consumption of highly chlorinated water). The problems with chlorine balance, said Osgood, were a result of the lack of training at the water purification stations.

> DAVID OSGOOD: And when you open the tap, you can probably smell the chlorine.

> TESSA FARMER: Yes.

> DO: So, you know that it's there. Years ago, I asked a guy—this was before I was working in water—and I asked him, "Is the water safe?" And he said, "Well, they chlorinate it enough that they probably kill everything." Now, these days you are hearing talk about chlorine being a carcinogen, so you are not exactly sure what the situation is. For instance, you know the Brita filters from the United States, so I brought one of those and I—it's a carbon filter and I'm not sure how [it's working]. I've operated on faith, I guess, that the filter is taking care of any problems.[10]

In Egypt, scandals about the quality of water are common. In 2010, drinking water from the Abu Ali water purification station in Kafr Magar, a village in the delta governorate of Kafr el-Sheikh, was declared unfit for human consumption (El Sherif 2010). Testing revealed high concentrations of ammonia, and there were reports of the spread of red worms and higher than normal algae concentrations. Despite long-running complaints by residents of the area about the smell and visible pollution, and despite reports from health authorities in Dessouq (released between October and December 2009) stating that the water was unfit for human consumption, no action was taken until a number of people who had become ill filed a complaint with the Dessouq Public Prosecution Office. A technical team created by the Dessouq Public Prosecution Office carried out water-quality tests and argued that the water must be improved. The Hisham Mubarak Law Center and the Association for Health and Environmental Development (AHED) also sent technical teams to investigate. The two sets of investigations seemed to have competing priorities. A physician in Kafr Magar, Dr. Shaimaa Ezzat Abdel Salam, had identified contaminated drinking water as the cause of severe intestinal complaints in young children she had seen, and a representative of the AHED accused the Dessouq Public Prosecution Office of intentionally intimidating the doctor, who was penalized for missing work on the day that she delivered the requested reports to the Dessouq prosecutors.

Water-quality horror stories came up in my conversations with Ezba residents. One afternoon, Asmaʾ and Teta told me a story about an incident in 2009 when drinking water was poisoned. Teta said, "In the countryside, a bunch of people died. It was reported on TV. This happened last year, didn't you hear about it? The government said to go buy bottled water. People said, 'Come on, let's open the tap and all die together!'" Asmaʾ replied, "I remember. People were telling each other at work, dads were calling home to their families to tell them not to drink the water [because] it has poison." "How did it resolve?" I asked them. Teta replied, "We were all sitting here for a while afraid. When something like this happens, what can we do? There was a report on TV that said that the water was safe again, so we drank it."

In August 2012, some 1,500 residents of the village of Sansaft in the Governorate of Monufia became very ill from contaminated drinking water that came from their home taps (Abdel-Baky 2012). Reportedly, the state-run treatment plant had pumped untreated water directly to residents' homes; the water had never received the necessary chlorine treatment due to staff turnover at the state purification station. The state station was one of two water purification stations in Sansaft; the other was unlicensed, created as a charity by wealthy families in the area. According to news reports, the village had long-term issues with the quality of the water from the state-run operation. Villagers mostly used the other station in town, but it had closed temporarily for the Eid holiday. It eventually came out that the state-run station had not conducted the mandatory tests in several years, despite the requirement to test the water every fifteen days. Water-quality testing showed the presence of salmonella and *E. coli*, along with high concentrations of heavy metals, including iron and manganese. The villagers revolted, holding the health minister and the governor of Monufia hostage for a brief period in the Menouf City Hospital. In response to this pressure, President Mohamed Morsi established a crisis management team and arrested the head of the station and some of his aides. During his brief tenure, Morsi attempted to create a Ministry of Water Supply and Sanitation Facilities as an oversight organization to manage the various entities that comprise the water and wastewater sector in Egypt, though this did not come to fruition.

A clip from the 2007 film *Morgan Ahmed Morgan* highlights this tension through humor, as the notable actor Adel Imam, portraying a parliamentary member, speaks to the Parliament against a variety of public complaints. About water and wastewater, this defender of the government's virtue says:

> These days, sewage water has contaminated potable water. This is immunization! This gives immunizations to the citizens of Egypt. Don't we give something to humans to make sure they don't get sick? This gives us immunizations, brothers! The people of Egypt have immunity now. And they say that "those who drink the Nile's water will be compelled to return to the Nile." Foreigners come and take immunization from us! What are you saying?! And then, where did this sewage water come from? Didn't it come from the Egyptian people? And it's going to the Egyptian people. [general laughter in the Parliament] (Idris 2007)

As these stories show, connections have various qualities. There are long-term and significant health impacts from these large-scale water problems, including diarrheal disease, which is responsible for so much childhood morbidity and mortality, and kidney and liver impacts from chlorine (Hamdy 2008). In 2010 the United Nations sent a special rapporteur, Catarina de Albuquerque, to evaluate the condition of water and sanitation in Egypt and to assess the extent to which the Egyptian state was securing these under the paradigm of human rights (Albuquerque 2010). The United Nations frames clean water as a human right, but what is clean enough? The answer is not universal, despite the notion that every human shares this same right. Water is local, and so is well-connectedness. We know that when we are foreign to a place—unconnected to it—we may not be able to safely drink the water. Being well connected to the water of a place depends on one's microbial environment and one's social interface with the water. I now examine two moments of trying to assess water quality, where people identified water as a potential vector for health problems but also saw it as embedded in larger contexts affecting their well-being.

Don't Drink the Water?

Water quality interested everyone in Ezba to whom I spoke about my research. There was often conjecture about what was in the Ezba water, but water quality was also contextualized in larger concerns about multiple vectors of contagion. Water is a hospitable place for nonhumans to dwell. In Ezba, I was interested in thinking through the many biological entities that converge around water in urban environments, and (more prosaically) I wanted to respond in a concrete way to the difficulties facing my interlocutors there. I therefore tried to find a way to get testing done on water from all kinds of places in Ezba, including faucets and various standing pools and puddles. My quest was ultimately unsuccessful. I had originally planned to work through Cairo University, but it fell through, and I was not able to make another arrangement in time. But the process of trying to get the water tested was itself instructive.

At Cairo University I had a contact through my Fulbright affiliation. Dr. Hani Moubasher was a considerate and intellectually generous research mentor while I was in Egypt, and on one of my visits to his office at the university, we discussed the possibility of testing:

> TESSA FARMER: I was also thinking about doing water testing, some on water puddles and [from taps]—
>
> MOUBASHER: You mean the puddles on the street?
>
> TF: Well, yes, because people are always complaining about water making them sick, so I wanted to see what was in it—from chlorine to [contaminants].
>
> MOUBASHER: No, that wouldn't work. You'd have everything in there. You'd have algae, small animals, insects, worms, bacteria, fungi, parasites. Look, this is Cairo University, best of the best, and you'd have to send the water to six different departments to get tests done on everything in there. It just wouldn't work. Without even looking, I could give you a list of what would be in there. Any standing water will have these things. Here, even in Cairo University, we had a water leak, and it left puddles. Within a couple of days, with the water in the sun, the green stuff which we call algae grew on top, then all of the secondary things came in, all of the other stuff that eats the algae and lives in the puddle. There's no point in testing that water.

A few colleagues then entered his office, and we were introduced. Moubasher appeared happy to be able to show off his Fulbright affiliation and his mentorship of an American doctoral student. The colleagues were involved in environmental research, and they started asking questions about my project. One asked how I had chosen Ezba. I told him that I had done research on water elsewhere through a fellowship and had enjoyed it, and I had then looked for a place in Cairo to do similar research. A friend working near Ezba had told me about their water problems and sent me articles on the blue jerrican demonstrations,[11] so I had gone there to check it out.

COLLEAGUE: That's very good, so you are the source of your own information on Ezba.

TF: Have you been?

COLLEAGUE: Me? No! I've only heard about it!

TF: What have you heard?

COLLEAGUE: I've heard that there are lots of, lots of, lots of—[making a piano-playing gesture with his hands]

Noticing my confused look and taking pity, Moubasher translated the gesture while copying the motion with his hands, saying that it meant that there are lots of small animals in Ezba. They all laughed. He told them that he had tried to assign one of his Egyptian graduate students to accompany me to Ezba on one of my previous visits. He had chosen the student as much for his imposing physical stature as for his knowledge of environmental science. But the student had called a friend who was a police officer, and the friend had advised him not to go if he could get out of it. "But," Moubasher related, "I told him, she's from America, and she's not afraid to go, and you as an Egyptian are afraid?" His colleague expressed his agreement with the student's reluctance to venture into Ezba, saying to me: "Yes, as an Egyptian I've heard about it but never been there. I'm surprised that you would go on your own." Water's valence reflects the reputation and the approachability of a place, and Ezba's water quality reflects its lack of well-connectedness to Cairo.

After his colleagues left, Moubasher and I returned to the idea of testing water in Ezba, given my interest in knowing more about water collected from kitchen taps. He agreed to provide water collection kits and gave me instructions on how to collect samples. On my next visit, I picked up the collection bottles and took them to Ezba.

I asked my contacts in the area if they were willing to let me test the water from their homes. Um Amina agreed and was very interested in finding out the results, as her children were often ill. When I delivered the water-testing bottle, she told me, "Oh, this will be useful for your research!" She was interested in the technical, scientific assessment of the water, but she primarily depended on her senses to determine whether water was safe to drink. She told me that for the last couple of weeks, while they were putting in the sewage pipes along Um Nadia Street, the water had been turned off to their area for two to four days at a time and then turned back on for a day or so. When the water returned, it stank of sewage for the first half a day. They would let the water faucets run until the water seemed to them to be potable again—until it smelled and looked fresh.

Um Amina used her sense of smell to alert her to problems and to decide when the water had returned to a sufficient level of cleanliness to be potable. This contamination had happened multiple times already, so she knew to leave the tap running to purge the unsafe water standing in the pipes. She combined these sensory processes with her knowledge of the pipe layout of her local area, which enabled her to predict which houses would be impacted each time. Her knowledge of local social connections helped her assess whom she could approach for help should she need to borrow water if the quality remained poor for too long a time. Um Amina said:

> We were very lucky on this street, as half the street [is connected by pipes to] Um Nadia [Street], and the other half [is connected] from up by the church, so only half the street was without water [this time]. People who had been cut off just took water from their neighbors. I just connected a hose to the house next to me and was able to get water without a problem. But the people farther down the street, it was very hard for them as they had to get jerricans and carry their water. And it [the outage] lasted for three days!

I also had a conversation with another resident of the area, Um Naser, who reported that when the water quality was poor, she made distinctions about what kinds of activities the tap water was fit for, evaluating the various levels of quality. Um Naser and other residents used water they judged unfit for drinking or cooking for other purposes: cleaning the home, washing clothes, and washing dishes. This in-between water was clean enough that it wouldn't leave clothing or the house smelling bad, though it was not at the level of safely consumable.

This shows that water quality is a spectrum, and local water is constantly moving up and down that spectrum with its current placement based on residents' judgment. People must be well connected with water in Ezba, for there is no guarantee that water from particular places or situations will be consistent in quality. Water quality is contingent over time, and water's contingencies shape the practices and rhythms of daily life in Ezba.

People were not passively responding to water quality, accepting whatever came from the tap; they actively sought to make their water more potable. Tap water for drinking was held to a different standard but still measured through smell and taste. Um Naser, for example, had a water filter in her kitchen tap. She used this filtered water to fill bottles, which she kept in the fridge. This water was not used for cleaning food or for cooking anything that would be boiled. Um Naser showed me her filter and complained that it was expensive, because the disposable inserts had to be replaced so often. Despite the expense, she was able to use it because she was married to a plumber and thus had easier access to filter replacements than did the average household in Ezba. As she pointed out, most houses didn't use water filters at all.

When Um Naser changed out the disposable filter, she said, it was dark and very dirty. Its condition made her wary of what was in her tap water even as the

Um Naser's water filter, 2010. Photo by author.

sewage system came online, supposedly ending problems with contamination. But even when there is no active sewage pollution, living nonhuman creatures are potentially present, as are nonbiological pollutants, both of which have the potential to negatively impact her family's health. Her wariness was increased by her repeated discoveries of a soap-like film in plastic containers that sat with stored water in them. "When I pour out the water [that has been sitting for two days] and I run my hand along the container, there is something that feels like soap. What is that?" The answer to her question is likely biofilm, a layer of microorganisms that are a widespread problem in potable water networks. (The film also was potentially an artifact of the container in which Um Naser stored the water.) Biofilms can contain microorganisms such as bacteria and fungi and higher organisms like nematodes, larvae, and crustacea (LeChevallier 1999).[12]

Residents of Ezba were very aware of water quality's relationship to health. This relationship was apparent to them through their own experiences with poor water and through frequent mentions by their doctors when they were ill. Many residents reported that their children were often on medication to clear up water-borne infections, such as parasites, amoebas, roundworms, and diarrhea. Several inhabitants reported that they were especially susceptible to hepatitis C (which they called virus C) because of poor-quality water. However, while hepatitis A and B are water-borne, hepatitis C is blood-borne. It is endemic in Egypt in part because syringes were reused during the mass schistosomiasis inoculations in the early twentieth century (Hamdy and Nye 2017; Ali 2002). Residents of Ezba said that they had heard that hepatitis C was water-borne from social connections who were diagnosed with the disease in pre-travel physicals. As this example shows, residents were very aware of the potential dangers of water-borne diseases, but they did not always have a clear picture of vectors of transmission.

Inhabitants of Ezba associated illnesses with things other than water quality, of course. But because of my own focus on water in the area and because the old traveler's adage of "don't drink the water" often rang in my mind, I usually attributed the many illnesses that I had during my fieldwork to the local water. During my conversation with Um Amina, I told her that I was being treated for an amoeba:

UM AMINA: Like worms?

TESSA FARMER: Something like that.

UM AMINA: The kids suffer from that. It's so frustrating. I've taken them to be treated, but it just comes back.

TF: Could it be from the water?

UM AMINA: Yes, it could be from anything. They eat out, whatever. Where did yours come from?

TF: I don't know.

UM AMINA: Do you eat in restaurants?

TF: Not a lot. Do you think that your kids got it from restaurants?

UM AMINA: Maybe. When one gets worms, they all have it—from washing clothes together, from using the same toilet. It makes me very mad.

Worms were a recurring problem in Ezba. In this same conversation, Um Amina told me that she had just taken her four children for medical checkups because they seemed so physically weak. The day before, she had asked her daughter to clean the stairs, but the girl had only done it lightly and then said that she was too exhausted to continue. The doctor diagnosed the three younger children with worms and prescribed deworming medicine for all four children, because if one had it, they all had it. Um Amina showed me the box of medicine. Four different kinds of worms were shown on the front cover, all looking fierce and disturbing, with lots of teeth. Um Amina laughed at my horrified reaction. From the same bag, she took other things she'd been given by the doctor: a huge bag of pills and liquid vitamins, which were supposed to improve the children's health. She had to physically force the kids to take the vitamins, running after them and holding their mouths open like cats being given pills, she said. When the children demanded money to buy the omnipresent goodies—ice cream, potato chips, *lib* (melon seeds)—she told them that these things were forbidden by the doctor as potentially the cause of their worms.

ELDEST DAUGHTER: Okay, I'll just buy the white *lib*. They don't have worms.

UM AMINA: No, the doctor said no *lib*.

YOUNGEST SON: I'm going to get ice cream!

UM AMINA: No, don't buy any [junk food]!

YOUNGEST SON: I'll just buy chips!

She continued to refuse, but the youngest son grabbed the money that he'd been given for successfully completing his exams and bought a bag of potato chips from a small snack store down the street. Um Amina sent her eldest daughter to take the chips away from him and give them back to the store. The kids kept going back and forth with her, and she continued to argue with them to keep them from buying the things that the doctor said would give them worms again.

I found it surprising that in Ezba, where residents had repeatedly been exposed to sewage in tap water, food was identified as the source of contamination. However, pulling apart the strands of causality for illness is not always easy, and there are many vectors for exposure to parasites, diseases, and toxins. The impacts of water systems are felt on bodies that are also implicated in multiple other systems in the city, including food preparation and distribution and medical care.

Ezba residents also complained about the sensory experiences caused by excess sewage—not only the smell but also tactile and aural experiences, for many mosquitoes and flies propagated in open sewage trenches. Um Naser attributed her son's rashy and infected skin to the overabundance of mosquitoes in her area, and she mentioned the frustration of constantly hearing the buzz of flies in the summertime. Of course, smell was the sensory experience that residents most often associated with wastewater challenges.

Um Samira, a young bride new to Ezbet Khairallah, lived in her husband's family's home because the young couple could no longer afford the apartment in another neighborhood that they had rented when they first married. Um Samira said that her daughter was born with allergies and illnesses due to the septic tanks. She never used to bring the baby outside, and she would often go and stay with her sister because of how dirty the tanks were, how much they stank, and how they made everyone ill. She was very happy that the sewage system had arrived and that the family had filled in their old septic tank. "The smell," she said, "made you ill even inside your own home."

As mentioned at the opening of this chapter, someone who was not from Ezba did not have the local knowledge needed to assess the water quality because they were not well connected *there*. Being well connected means having the capacity to discern what is well connected enough, what is good enough for various purposes. Being well connected means having to and being able to manage the subtleties of your own senses in order to determine what is good enough. Indeed, in Ezba, being connected well was not always good at all. In Ezba, even the new water and sewer connections were not necessarily *good* connections, for residents still imbibed water of questionable quality with potentially harmful effects. The doctor's complicated instructions about what Um Amina's kids could and could not eat was actually another iteration of the project of becoming well connected. In Ezba, well-connectedness to home was an ambivalent status, one that you had to seek to survive (see Rubaii, forthcoming, on ingesting cement dust as placemaking). In Ezba, sensing connection was about using one's senses as instruments of learning and engaging with an imperfect world to make life possible and to connect well enough to the place one called home.

What Tastes like Home?

Sensing connection is a complex and nonbinary project. The overlapping qualities of sensing—sensory density and acuity—allowed residents to manage the nuances inherent in the spectrum of water quality in Ezba. The trained body, porous and perceptive, was a tool for diagnosing quality—an instrument that had been socially conditioned through certain kinds of local interpretive practices. An extended conversation in 2018 with Sara, a young woman from Ezba, laid out this embodied experience:

> TESSA FARMER: Other people used to tell me that when the water cut, sewage water would get into the pipes.
>
> SARA: Yes, I've heard the same thing, that sewage water was getting in the pipes. People say that. Maybe that was what it was when I thought that the water was dirty at that time.
>
> TF: People told me that you could tell the level of cleanliness by the smell.
>
> SARA: Yes, the smell [of bad water] was clear. Now the problem is that there's too much chlorine in the water.
>
> TF: Yes, and that's bad for your body.
>
> SARA: Of course, it [alters] the liver and kidney. It affects all of that. This is one of the things that I smell sometimes. I smell chlorine. Sometimes the water is totally ordinary. Sometimes when I go to smell its scent, I feel that the water has chlorine in it. I feel like that. When I go to drink the water, I feel that the flavor has something in it. Sometimes the water is good, it doesn't have anything.
>
> TF: Sometimes this, sometimes that.
>
> SARA: Not always, honestly. It's not all of the time that the water isn't clean. A lot of the time the water is clean and doesn't have anything in it. And sometimes from the smell you think that the water has too much chlorine in it.

The spectrum of water quality that Sara articulated, which ranged from too dirty to too chlorinated to just right (water with no particular olfactory qualities), required constant processes of sensory assessment. The safe levels of chlorine were much murkier and more difficult to sense than those of sewage because residents could easily tell when the sewage admixture was bad; this was what their senses had been trained to identify. Part of the murkiness was that chlorinated water is "clean" but not necessarily healthy. As Sara described her processes of managing overchlorinated water, I recognized them as the same practices she used for water contaminated with sewage:

> TF: What do you do when the water is that way [has too much chlorine]?
>
> SARA: Often, I don't drink from it. I leave the faucet open and wait a bit. We have the expectation that if a lot of water runs, it will finish with this and better water will come. I guess! [laughter] Some people leave water out, because chlorine will dissipate. If you leave a cup of water for a little while, it [chlorine] flies away. Some people say to boil the water [when there is sewage]. You get a very big pot and leave the water to boil on the flame. After it cools off completely, then fill bottles and put those in the refrigerator. Don't drink water from the tap. But who is going to be free enough to do that every time? There is laziness. There are going to be people who don't want to wait for it to boil, then wait for it to cool down, then put it in bottles and put them in the fridge, and then wait until it's gotten really cold just to drink. Right?
>
> TF: Right.
>
> SARA: It's a difficult issue.

Sara demonstrated the management practices she used for water she sensed as not good enough. For both problems, Sara said, time was key: for sewage, water was left running from the tap over time or was boiled and then cooled, and for chlorination, water was left to sit in open containers or the tap was again left to run.

Sewage admixture was a problem at many scales. Sometimes, when a whole area had no water pressure, sewage got into the pipes. In this situation, the neighbor's water was as bad as my interlocutor's, and everyone just had to wait while the water ran. Other times, my interviewee's own pipes allowed sewage to leak through the joints. This issue couldn't be resolved by waiting, but neighbors might have clean water for them to borrow while they got their plumbing fixed. Sometimes people didn't know where the sewage originated. In those cases, time and connection again were the answers. Residents tapped their social networks to learn how widespread the problem was, and by process of elimination they could then pinpoint the source of the contamination. Social relationships and local knowledge were thus central to the project of sensing and managing water quality.

What is "good enough" has changed over time and across space.[13] As the state-run wastewater system came online in Ezba, the incidents of sewage admixture dropped dramatically, though they did not disappear completely. Water that was of bad quality because of sewage contamination was localized, but water of bad quality from chlorination remained systemic. As Sara mentioned, managing the problem of overchlorination required time, which she and others did not have. Some residents, like Um Naser, installed a water filter on their kitchen sink to filter drinking water.[14] When I asked Sara about the filters, however, she brought up the issue of taste:

TF: There are filters.

SARA: Yes, I will tell you something strange. When I drink filtered water, I can't finish it.

TF: Why?

SARA: I don't like the taste.

TF: Why?

SARA: I don't know. My uncles got filters at their houses. Once when I first got to their house and said that I wanted to drink water, my aunt said, "Drink. Put the filter and drink." The faucet on the filter is small like this. [gesturing] When I went to drink, I said, "The taste is awful!" She said, "Oh, you are just used to dirty water. Go and drink from it!" [laughter] I told her, "The taste is strange, I swear to God!" Maybe it's because I'm not used to it. Maybe. Water that is very clean, the taste is pristine. The water from the faucet, you feel like there is chlorine. I don't know. The issue is that they are different. The taste is different. The taste of this [faucet water] and this [filtered water] are different.

TF: How would you compare water from the faucet to bottled water?

SARA: Bottled water? I don't know. When I go to drink from bottled water, I don't get satisfied. I swear to God, I don't know why. I feel like I didn't drink anything. And when I drink from the faucet, I feel like I drank. [laughter] I don't know why.

> Maybe because we are used to it. Maybe it's the idea of being used to something. When I'm outside in the streets for a long time, like when I go to the university, I will buy water. A bottle of water. . . . A few times I have bought a bottle of water and felt like it was sour. Have you experienced that? Did it happen to you once, for example? When you get bottled water, it's a strange taste. Right?

In Ezba, residents often solve contamination by displacing, dispersing, or disposing of the bad water; part of being well connected is the capacity to move things away from you when necessary. With water, this process of displacement may be boiling and cooling or letting the pipes run. Part of making a place one's home is dispelling the things that make it inhospitable: dirt on the floor or on clothes, worms in the children, waste in the water. To stay in place, Ezba residents seek to dispel contamination and displace objects that are toxic, and to do so they must sense and engage the undesirable. But to stay in place—to stay at home, where undesirable elements are part of life—people must also consume them. On the one hand, they want to remove toxins and other contaminants; on the other hand, as Sara's discussion of the taste of water shows, getting rid of too many contaminants is itself alienating—the water does not taste like home. To Sara, filtered water, bottled water, and water from other areas of Cairo all taste strange. Satiation, for Sara, is about the ability to pick up enough sensory cues to recognize specific water as "home," and water at home in Ezba simply has extra things in it at times. Living with the problems is part of being at home.

The quality of water is not a binary—good or bad—as my friend who stepped in the puddle saw it. In Ezba, water quality is on a spectrum. It is, of course, about health and sickness, but it is also about meaning and value. Sensing and qualifying water are part of the process of connecting well.

Sensory Interpretation

One morning, I was riding in a shared taxi into Ezbet Khairallah. The driver stopped at the top of the hill above the el-Zahraa metro stop to pick up a woman and her small child. After climbing into the car, the woman told the driver, "Oh, I can't stand the dirt and filth and garbage and smell"—as if by way of explanation for requesting to be picked up at what would normally be halfway through the trip. In response, the driver replied, "Ezba is full of it!"

> WOMAN: Ezba is covered in garbage.
>
> DRIVER: Here, if people don't smell garbage in the morning, they get allergies.

WOMAN: It smells terrible.

DRIVER: If there isn't garbage in front of the houses, they'll call the Zabbaleens [garbage collectors] and tell them, "Leave me a bag!"

WOMAN: The place is a good one, but the people are dirty. It could be nice if the people cared.

DRIVER: The people here are filthy.

WOMAN: Well, there are some good people here too.

DRIVER: No, finding good people in Ezba is impossible.

The garbage, the sewage, and the scarred and dirty bodies are what people outside Ezbet Khairallah see. These refrains echo through newspapers and popular media, and they appear over and over in conversations with people living nearby. Over the last century, Ezba has been used as a wilderness retreat, a garbage dump, and a cemetery. Since the 1970s, it has reinvented itself again, this time as a densely populated informal settlement. But Ezba has always been the kind of place people don't go unless they must.

These stereotypes aren't entirely inaccurate, but they are incomplete. On my first visit to Ezba during daylight, garbage and sewage did indeed greet me. Ezba is built along the slope of a limestone ridge, and shit literally rolled downhill from burst pipes and overflowing septic tanks, and it was mingled with water discarded from buckets at doorsteps. The waste followed the oddly angled, narrow streets to the bottom of the hill where the microbuses linking Ezbet Khairallah to Fustat, el-Malek el-Saleh, Sayyida ʿA'isha, and Khalifa stop. There were garbage piles along each of the major entrances to Ezba, heaps of plastic bags and organic waste playing host to rodents, semi-feral cats and dogs, and even occasional herds of grazing sheep and goats.

This refrain of Ezba-as-sacrifice-zone fails to account for the settlement's economic vitality and its upward trajectory. During my fieldwork, Ezba was gaining ground, and this was visible in the construction of the state-run sewage system, which broke open streets, puncturing them with gaping trenches dug by backhoes or by hand. In some senses, Ezbet Khairallah was in a tenuous and hopeful stage of upward mobility, offering residents hope for a better future as it gained new services and increasing recognition from different state entities. However, this hope was mixed with suspicion, just as the water was mixed with sewage. Residents wondered about the real goals of these improvements, and they felt anxious about what the improvements would mean for them.

Sensing connection in Ezba means many things, including assessing thresholds between potable water and wastewater, between social connection and social isolation, between the inside and outside of bodies. Where does this particular water source at this particular moment fall on the continuum between sustaining and threatening? Sensing connection is about assessing not only the quality of water but also the quality of connection. How well connected is Ezba with the mechanisms of municipal governance? How well connected are Ezba's people to those in other regions around the country and in other neighborhoods near them? Residents had to tap into networks and connections to locate the necessary ingredients for life in this space. Sometimes, water came in from el-Basateen on heads. Virtual water came in the form of tomatoes, arugula, and the *baladi* chicken that arrived with a nephew from the country who sought a job in the city. Residents tested for the smell of good and bad connections alike. They measured the weight of conflicting impetuses. They sensed and hedged their way into practices for living amid the intersecting tensions of relationships that require mutuality. They experienced the limitations and possibilities of Ezba in their bodies.

In the story of water systems in Ezba, bodies mattered. They carried pails and pushed puddles of sewage. Bodies bore the signs of making do: feet and kidneys and livers were scarred; a young girl's eyeballs were deformed from the labor of carrying water up and down five flights of stairs every day on her head. If one looks closely enough, however, these bodily traces of making do may also index thriving. The bodies of Ezba residents were enmeshed in the water and sewage systems that they engaged in. These systems coalesced in moments when somebody asked for an accounting of what life was like in this place, and the answers included concrete steps leading up from flooded streets, brick walls in living rooms that collapsed after sewage water corroded their foundations, and PVC piping in a neighbor's house that broke and flooded a home in minutes, destroying twenty years of accumulated bridal china months before a wedding.

Residents bore the weight of water and felt the pressure of it as it poured or trickled through pipes (Anand 2011). Water mediated social expectations and political possibilities as these intersected with bodies and neighborhoods, which residents were always working to build. These experiences were mediated by senses beyond taste or smell or sight.[15] The sense of pressure is called proprioception. The vestibular sense located the residents in space, both as individual bodies and as a communal body living in this particular neighborhood—a neighborhood whose value could be indexed by its quality of water. Then there was interoception—an awareness of the bodily responses to affective states. This sense mapped from within themselves the pain and possibility of the social worlds that were made possible through the material, moral, and ethical connections that they worked to build in and through water. These three senses, along with sight, smell, taste, touch, and hearing, were mechanisms through which residents as individuals and as members of social networks experienced their world. They obtained information, assessed potability, tested for possibility, recognized limitations and thresholds, and gauged the possibility of making connection happen well. Using these socially and bodily trained sensory processes, residents asked: can I find enough of the right kind of water at the right time to make life possible? How do I locate myself as an individual within the social, ethical, and material goods and ills of connecting in Ezbet Khairallah?

Chapter Six

Conclusion

Gathering Connection

The proprietor of a small in-home snack store in Ezbet Khairallah had a white plastic bucket sitting on the doorstep of her house, which she kept filled with water from her bathroom tap. I had noticed it when walking in and out of her store before, but I hadn't remarked on it. One day while we were sitting in her customary location just inside the entranceway to her house, a dog came up, looked around carefully, and took a drink out of the bucket. I turned in surprise to Um Nahla, and she smiled: "That's what it's there for." I asked, "You have a *sabīl* for dogs?" I was surprised, as I had only previously seen *sabīls* (charitable water fountains) for humans. Her comment made me realize that I had missed an additional layer and an alternative material form of *sabīls*, which are ubiquitous sites of social, material, and ethical connection in Ezba.

UM NAHLA: Dogs, cats, and birds.

TESSA FARMER: Really? I never heard of a *sabīl* for dogs and cats. Maybe birds.

UM NAHLA: Weren't we told to take care of everything with a wet liver?

Um Marwa was referring to a *ḥadīth* in which someone asked the Prophet Muhammad if one would earn rewards from God for the care of livestock. Not just livestock, he replied, but anything with a moist liver. The equivalent in English might be anything with a heartbeat. Um Nahla said that the idea had come to her after watching a cat come in to drink out of the bucket intended for watering the ducks and geese living under her stairwell, which were gifts from her daughter's fiancé. Um Marwa was a longtime widow with a number of children, whom she supported on her husband's death benefits plus the modest income from the sale of cookies and soda at her shop. For her, any kind of larger *sabīl* seemed out of reach. "Do you gather *ḥasanāt* for a *sabīl* for animals?" I asked. Um Marwa replied, "Of course. A man went to heaven for giving a dog water out of his shoe."[1] This moment highlighted for me the shifting evaluations of what is possible in Ezba. Nigel Thrift (2004) might call it the "politics of anticipation," where the future is uncertain but contains the possibility of better things—both in the material sense (improved infrastructure) and in a local cosmological sense (preparations for the afterlife).[2]

In Ezbet Khairallah, *sabīls* are a way of sharing water with neighbors and passersby. Creating and maintaining a *sabīl* gathers merits with God in the Islamic tradition. Those merits (*ḥasanāt*) can be shared by all who help to maintain the *sabīl*, and they are often passed on to a departed loved one through a dedication. *Sabīls* are unique sociomaterial objects in which the themes of my research come together. *Sabīls* condense within themselves the diversity of material forms and practices of care and repair for clay and metal water infrastructure; the embodied notions of smell, taste, and temperature; a shifting history of social responses to a material context of hardship; and practices of neighborliness that draw on religious traditions to shape the livability and traversability of this neighborhood.

Sabīls embody what it means to be well connected in Cairo. Like a rainwater barrel gathering water, a *sabīl* is a mechanism for gathering connection. As I have discussed, being well connected in Cairo is a lifelong project that operates at social, material, and ethical levels as residents build and navigate networks: interpersonal networks among themselves and broader networks linking them to the neighborhood that they are sowing into being. As Um Marwa's *sabīl* illustrates, the neighborhood is not limited to human occupants; it also includes the dogs, cats, and birds that drink from her *sabīl*. The neighborhood is not even limited solely to corporeal beings, for it includes the souls of departed loved ones and the host of jinn and demons, all of which coexist in Ezbet Khairallah. Being well connected requires astute social skills to manage the benefits and pitfalls of exchanging, valuing, and evaluating potable water and the connections it fosters. *Sabīls* exemplify this locally specific form of the sociality of water, which requires embodied practices and habits of social relations to make good connections feasible.

Water is both a product of and productive of certain geographies of difference. It maps colonial legacies and neoliberal agendas in space and on people's bodies. It is a medium for kindness, love, and connection. It also marks Ezba and its inhabitants as deficient and dangerous, as needing management and sanitation—indeed, sanitization—to contain its hazards, to protect the rest of the population from the disease, rockslides, theft, and prostitution that might travel outward like contagion. Like a modernist duality symbolic of the city writ large, water is both citizen maker and source of pestilence, and like the city, it is intimately involved in the flows and frictions of capital in this structurally adjusting state. Connecting well is a frame that offers a closeup view of systems in place: how people craft connections and fit them into systems and spaces, and how the networks of connection show what it is to be a human engaged in relations with human, animal, material, and spiritual counterparts.

Moralities of Connecting Well through Water

Sabīls offer the possibility of ethical connectivity. A *sabīl* gathers in material form the connections among neighbors; it gathers *ḥasanāt* and stakes the moral claim that the people of Ezba are good residents. When I spoke with various *sabīl* keepers in Ezba, they cited four main interrelated aims for creating and maintaining their *sabīls*: to be good in a religious sense, to work toward being kind in their everyday lives, to manage grief in the case of a loss in the family, and to solve practical problems.[3]

People who created *sabīls* to be good in a religious sense sought to establish an object that would enact the Prophet's religious exhortation to share water with every creature that has a moist liver. In Islamic theology and local interpretive processes, the sharing of water is highly valued. Water is seen as a key element of the composition of all living things, and it is therefore the crucial component of God's divine creation, drawing on *sūra* al-Anbiyāʾ (21): "And We made every living thing of water." Water is powerful in the imagery and everyday practices of spirituality, being integral to material and spiritual purification. The importance and value of sharing water are reiterated in a number of *sūras* and *ḥadīth*, and individuals in Ezba drew on these ideas in their decisions to set up *sabīls*. *Sabīls* can be created as acts of *khayr* (good deeds) or *ṣadaqa jāriyya* (perpetual charity).[4] For those *sabīl* caretakers who sought to be good, creating and maintaining the *sabīl* was a religious act, which also had the possibility of spiritual benefits for the souls of their deceased loved ones.

Saleh Lamei Mostafa (1989: 34) defines *sabīl* as "a road, or a path," coming from the verb *sabala*, "to let fall, drop, to let hang down, to close eyes or to shed tears." Mostafa argues that the use of the term *sabīl* to refer to a building created for a public drinking fountain likely comes from its connotation of being a work done on behalf of God, *sabīl li-llāh*. The association with a road was augmented in Ezba because it was a gesture intended for travelers or passersby. A *sabīl* is a gesture toward the afterlife (discussed further below) but also a commitment to the well-being of one's current community—"a bodily need met with a spiritual response," as one resident of the area told me.

Being good was thus deeply tied to the next motivation: to be kind. The maintenance of a *sabīl* makes God's goodness manifest in the world in its creator's intention to benefit the material recipients of the water. *Sabīl* proprietors often said that the desire to extend kindness was central to their mission, and they highlighted several categories of intended human recipients, including neighbors, children, people taking pills, and those walking long distances. In this, residents recognized the vulnerability of particular bodies and the difficulty of obtaining water there; they were thus in part responding to their sense that Ezba was a particularly difficult place to traverse due to its topography and the rough state of the roads.

Of course, *sabīls* are not the only way to be good or kind. When I asked about the larger context of options for charity, residents offered a range of acts, from a simple smile up to grand gestures of dutiful obedience to the will of God (see Moumtaz 2021 and Mittermaier 2019). Residents also often cited the distribution of copies of the Qur'an or the purchase of shrouds (*kafan*) for people who couldn't afford them.

The third motivation for maintaining *sabīls* directed the first two impulses—goodness and kindness—in a particular way: through dedications. *Sabīls* dedicated to dead loved ones were vernacular memorials, allowing mourners to grieve their losses. They also served the dead more specifically. A number of *sabīls* included messages asking users to carry out particular religious acts as part of the consumption of the water. The two most common requests were for the user to perform a *duʿāʾ* (prayer of supplication) or to recite the opening verse of the Qur'an, *al-Fātiḥa*, on behalf of the deceased to whom the *sabīl* was dedicated.[5] Users reported offering *duʿāʾ* such as "May God have mercy on this soul" or "May he drink from the rivers of paradise."

Even without signs requesting particular kinds of prayers, there was an implicit expectation that offering *duʿāʾ* was a good thing to do as a *sabīl* user.[6] The signs simply made it clear that a particular *sabīl* was intended as *ṣadaqa jāriyya* (per-

Sabīl of ʿ*ulal* with dedication, 2018. Photo by author.

petual charity) with a dedicated beneficiary. Even without the prayers, the interaction with a *sabīl* accrued benefits both for the drinker (for the act of consuming the water) and for the proprietor (for the ongoing act of charity). Indeed, drinking from a *sabīl* was as much a form of connection as maintaining one. Some *sabīl* users chose to drink from a particular neighbor's *sabīl* as a way to advance the neighbor's project of *khayr* or *ṣadaqa jāriyya*, and others said that they avoided some *sabīls* precisely because they did *not* want to support the project of *khayr* of those whom they disliked—a form of hedging connection. The material forms of these connective objects laid out differing possibilities for well-being, differing levels of commitment to maintenance and repair, and differing sensorial potentials for *sabīl* users.

The fourth category of reasons that people cited for their decision to establish a *sabīl* revolved around practical, prosaic considerations. The offering of water helped them to be well connected with their neighbors. For example, shopkeepers were often asked to provide water to customers and passersby in the everyday course of their work. Many of the small storefronts carved out of apartment buildings did not include water taps, so shopkeepers brought in water every day. *Kūlmān*-style *sabīls* were often a solution to these requests.[7]

What Makes Something a *Sabīl*?

Standing on the corner of two small streets on a hot summer afternoon in Ezbet Khairallah, I could see examples of all four main types of modern urban *sabīls* in Egypt.[8] Both of these streets were unpaved and meandering residential lanes in the unplanned settlement. The buildings at this small intersection were of redbrick and concrete, built cheek to cheek and three to six stories high. Apartments occupied the majority of the space, and many had a small storefront on the ground floor.

Entering one of the streets after walking under the ring road that cuts through Ezba, I saw to the right a series of four small earthenware pots called *ʿullal* (sing. *ʿulal*), which were arranged on a wooden shelf some five feet up a wall. The *ʿullal* were a deep red clay with simple white wave patterns. They had rounded bottom chambers and narrow necks covered with earthenware lids. Even on this hot day in June, they were cool to the touch. *ʿUllal* work through evaporation. Warm water percolates outward through the clay chamber, and heat is drawn from the water within, which propels the evaporation of water from the outer surface.[9] Each of these particular *ʿullal* had a built-in earthenware filter at the base of the neck to prevent foreign materials from dropping into the water. Bringing one of these *ʿullal*

(Top left) ʿUllal, 2018. Photo by author. *(Top right) Zīr*, 2018. Photo by author. *(Bottom left) Kūlmān*, 2018. Photo by author. *(Bottom right) Kūldīr*, 2018. Photo by author.

down from the wooden shelf, I could smell deep earthy notes with a sweet undertone. The taste and smell of *ʿulal* water were described by several people in Ezbet Khairallah as incommunicable, unique unto itself but essentially pleasant. To me, they smelled of warm brick. These four *ʿullal* sat outside a woodworking shop and were cared for by the shop's owner as a form of *khayr. ʿUllal* are the cheapest of the *sabīl* options, costing between EGP 7 and EGP 15 each based on the quality of their construction. They are easily replaceable.

The next building, closer to the intersection, had a light brown *zīr*, damp around the rim. *Zīrs* operate on the same evaporative principle as *ʿullal* but are much larger, with a barrel body and a pointed bottom. At around EGP 70, *zīrs* are a more expensive option, but like *ʿullal*, they are used until broken, then replaced.[10] This *zīr* sat in a wooden frame built from scrap lumber and was cushioned by an old bicycle tire. A small plastic cup, for scooping and drinking the water from the *sabīl*, sat atop a repurposed water barrel lid. The flavor of this water was much like that from an *ʿulal*, though some *zīr* proprietors added rose water, cut fruit, or fresh mint to their *sabīls* to improve its taste and scent on the principle that in charity, we should give the best we have—what we ourselves would like to receive.[11] This particular *zīr* used to be located in the entryway of the home, where the family stored drinking water as they had done in Upper Egypt before moving to the area. The proprietor had moved the *zīr* onto the street just outside his door, transforming it into a *sabīl* in honor of his mother when she passed away some three years prior.

Nearby stood a medium-size *kūlmān*, an insulated plastic jug with a spout, which was set atop a folding stool just outside a small market store.[12] *Kūlmāns* offer colder water. Ice, most often in the form of a frozen bottle of water, was added along with the fresh water each day. *Kūlmāns* were moderately expensive at EGP 90–150, but were a common household item. This form of *sabīl* was the one most commonly added during the hot summer of 2018 and was the kind most likely to be transitioned back and forth between personal and public use.

Finally, kitty-corner to the *kūlmān* and several buildings down the cross street, sat a mechanized water cooler, a *kūldīr.*[13] *Kūldīrs* are the only form of *sabīl* that must be connected to water and electricity, and they are by far the most expensive option—some EGP 1,200 for the unit. While this one was not, many *kūldīrs* were encased in a metal frame to prevent the theft of parts and damage by passing cars. This type of refrigerated *sabīl* was often used by residents to refill plastic water bottles. (See Kaplan 2011 on the social relevance of water coolers.) This particular unit had been in place for about seven years, appearing just as the area transitioned to the piped sewage system, and it was constantly under repair in 2018.

The temperature of the water from earthenware pitchers, both *zīrs* and *ʿullal*, is moderately cold. The water in non-earthenware *sabīls*, both plastic (*kūlmān*) and metal refrigeration units (*kūldīr*), is much colder. A number of *sabīl* users said that they preferred drinking from earthenware *sabīls* because of the temperature, taste, and scent. The clay itself was also believed to pull impurities from the water, offering a healthier water source.

The form, history, and context of the four *sabīls* at this intersection in Ezbet Khairallah demonstrate the material, ethical, and social relationships that are built through the project of connecting well. The vernacular *sabīls* described here build on a Cairo tradition. There are historic buildings, primarily established during the Mamluk and Ottoman periods, with *sabīls* that are colocated with *kuttāb* (centers for Islamic education; Moumtaz 2021; Mittermaier 2019). *Sabīls* are important parts of the built environment of Cairo. They are often located in areas underserved by municipal water systems and are particularly significant given Cairo's changing environmental conditions. As these *sabīls* show, vernacular and small-scale water infrastructure can add to urban water adaptation; however, small-scale water sharing is often overlooked in resource management, which focuses primarily on large-scale issues and state discourses (Leeuwen et al. 2019; Pahl-Wostl et al. 2007). More broadly, public fountains have historically been a part of many urban traditions (Mostafa 1989), and Wutich and colleagues (2018) argue that water sharing at a household level is a cross-cultural phenomenon in which each location has its own constellation of ethical frameworks, technologies, and ontologies (see also Brewis et al. 2019). The forms of the *sabīl* that I examine here grow out of the Islamic tradition that water is divinely provided and local conceptions about the importance of sharing water in difficult urban contexts.

Residents of Ezbet Khairallah reported that there had been few *sabīls* in Ezba until around the turn of the twenty-first century, when the settlement informally connected pipes to municipal water supplies. Few people could afford the time or effort of hand-carrying water from neighboring areas, nor would they purchase it from the infrequent rounds of informal water sellers only to give it away.[14] However, it was possible during that time for residents to knock on a door and ask for a glass of water from a neighbor or from a stranger's house as they passed through the area. As Ezba developed and self-made water connections proliferated, *sabīls* began to be established, at first mostly at mosques and in front of businesses. Another small wave of *sabīl* creation happened in the mid-2010s after the GOC installed the sewage network, reducing contamination between potable and black water.

In my conversations with residents in the summer of 2018, they noted that in the previous two years, the number of *sabīls* in the area had multiplied dramatically. The creators of several of these new *sabīls* told me stories of deaths from heat exhaustion and of their own experiences of thirst as they moved through the urban landscape. Others said that they had established *sabīls* because of particular environmental conditions: their street was particularly hot (unshaded), long, busy (a main thoroughfare or a location with many workshops), or dry (without other *sabīls* nearby). Still other *sabīl* keepers were responding to the loss of a loved one. A *zīr* or *kūlmān* might have transitioned from personal use to *sabīl*, proprietors might have gone from occasionally providing water to establishing a dedicated *sabīl*, or an existing *sabīl* might have been upgraded from one form to another (e.g., *ʿulal* to *kūlmān*). The proliferation of *sabīls* was in part seasonal responses to climate change and to the fasting month of Ramadan, which fell during the hot period. These temporally induced *sabīls* added to the already existing network of charitable drinking fountains, which had been created as a local response to urban water scarcity. Residents of Ezba have long found ways to share their precious water as they have transitioned through the various arduous systems for obtaining it.

By establishing these charitable water fountains, residents of Ezbet Khairallah sowed connection: they took an ethical approach to their neighbors' water needs and at the same time laid the foundations for longer-term cosmological well-being for themselves and their loved ones. They located connections through *sabīls* placed at the doorsteps of homes and in front of small stores, offering water near where they lived and worked, creating asynchronous connections with passersby. Residents also hedged connection through *sabīls*, choosing which water fountains to use based on their differing investments in the ethical well-being of particular neighbors. They also used *sabīls* to manage the practical challenges of operating small stores, deploying *sabīls* to reduce the burden of providing water from home. Residents sensed connection as they drank from *sabīls*, assessing the smell, taste, and temperature of the water as well as the visual appeal of the fountain itself; this mirrored the processes by which they evaluated the quality of water in the rest of their lives.

Sabīls illuminate a long-standing and undervalued set of practices that can inform those of us with easy-running taps of what is at stake in water—not only the condition of water itself, but the social entanglements that material difficulties and opportunities enable. *Sabīls* gather together all of these themes, just as they gather *ḥasanāt* and connection. At the *sabīl*, residents of the area make connection happen well through water.

Well Connected through Water

Residents of Ezba live in difficult economic, social, physical, and political circumstances. The systems and networks within which they live are characterized by both flows and frictions, and they bring dangers and punishments, joys and flourishings both broad and incredibly intimate. The questions of how people make lives, make livings, make homes, make meanings, and make communities (which they also destroy and undo) have long been central concerns of social sciences. As I have shown, through everyday forms of collaboration and evasion among neighbors and with representatives of state institutions, people in Ezba strive to make water and wastewater utilities work for them. I have documented not the Egyptian state in the abstract but as it is visible through the water sector—the always-in-flux institutional and organizational paradigms that have sought to manage and dictate relations between water and people in Egypt, and the intracommunal networks necessary for making water move or stay in the right amount, with the right quality, at the right times. I have looked closely at water in a particular informal settlement to show how water matters in people's lives—as a vital resource and a basic right—as it is conceptualized in Egypt.

A picture has emerged of the social and political ramifications of the material forms in which water is available—or not available. The Nile River, as the primary water source for the country, has deeply shaped the lifeways of the Egyptian people, the country's political structures, and its geopolitical relations with upstream neighbors. The task of managing this water (or attempting to do so) has underpinned both colonial administrations and the modern Egyptian nation-state. Institutions that were created to professionalize and operationalize water management at the national and local levels have evolved over time in relation to changing political and economic factors, perhaps nowhere more so than in the capital city of Cairo.

The city of Cairo has a long history stretching back over a thousand years, and the proximity of Ezbet Khairallah to the remnants of this history is part of the reason the area acquired its state-run sewage system. The water coming down the limestone hill from Ezba's unlined septic tanks had threatened to dissolve the artifacts buried in nearby Fustat. From the early days, Ezba has faced the danger of being razed; because it is an informal settlement—*ashwā'iyyāt*—residents are legally defined as squatting on state-owned land. From the ISDF to the Cairo 2050 plan, it's clear that many powerful entities have plans for the area that do not include the current residents.

The affective connotation of living in an *'ashwā'iyyāt* is a lingering challenge for Ezba's residents. Although more than half of Cairo's population live in places that fall into the category of informal settlements, certain places are considered more hazardous than others. Ezba has been marked as particularly dangerous. The miasma of its reputation could for many years be felt as a sensory punch in the nose when entering the area. The ever-present sewage marked spaces and bodies as being in particular relationships—either inclusion or exclusion, or sometimes both—with the city and the welfare functions of the state. Ezba's overflowing sewage was produced by the colonial legacies and neoliberal agendas that have shaped so much of the material base of Cairo. It physically manifests social and political relations, while at the same time it shapes those relations as they move forward through time. In the dual imaginary of the "modern," the city is where people are remade into "proper citizens," but it also produces pestilential dangers. Sewage mixes these meanings: it is the excess and mobile product of unmanageable populations that threaten the well-being of those who have been properly incorporated into the polity. Sewage from Ezbet Khairallah must be managed to mitigate the threats it poses to Cairo proper, to the valuable remains in nearby Fustat, and even to Egypt's international reputation. The government does not want another public relations disaster like the 2008 el-Duwayqa rockslide.

The sewage system that was installed in Ezba between 2009 and 2011 offered the possibility of solidifying residents' claims to ownership, creating a grid of officiality in utility bills and measured housing plots. However, the sewer system and the state recognition it offers are proving to be a double-edged sword. The improvements have pushed up rent prices, which are now out of reach for the poorest section of the population. The recognition from the state has also raised the specter of dispossession, given Egypt's history of forcibly relocating poor urban populations. The sense of possibility—the vision of a revitalized and improved Ezba—is matched by visions of dystopic futures in which Ezba's new services and infrastructure become the basis for a wholly different Ezba, given over to those with *wāsṭa* (connections, nepotism). As residents of Ezbet Khairallah well know, in Egypt, *wāsṭa* is as common as zucchini. They also know, perhaps better than anyone, that good fortune and progress are laced with danger, just as water may be tainted with sewage.

The state had long ignored informal areas like Ezba, allowing them to exist without formally recognizing them by installing official service connections. This "seeing without seeing" echoed the practice of *istiʿbāt* (deliberate misunderstanding or misrecognition)—a technique of hedging that has shaped interactions between Ezba residents and between them and state institutions and their representatives. Through this sort of hedging, parties manage their connection with each other in ways that would not be possible with complete directness and transparency. For example, inhabitants of Ezba do not have formal tenure rights to their land, but they are able to register for food subsidies based on their residence there, and they are able to vote for candidates specifically elected for the area in which they live. Representative Magdy Allam, who first called this "seeing without seeing," implied that when overeager civil servants try to actually bring places like Ezba into state view, they put themselves at risk for having disrupted the façade. Similarly, people selectively engage and evade state involvement. While Ezba residents were keen to have the state sewage system and worked to make it happen, the new system also introduced new modes of surveillance. Residents worried that the state would monitor what goes down the drain, for the conversion of one utility into an official system brought other informal arrangements into view. Residents wanted the state to see them, but not to observe too closely.

After the rockslide disaster in el-Duwayqa in 2008, Egypt's visual acuity improved. Through connection (both official, like Ezba's connection to the city sewer system, and unofficial, like neighbors who came together to agitate for the services), area residents have staked a claim to their right to the city. This is a two-way street in which the governorate has become responsible for providing services to a greater number of locations, but residents of these locations then become responsible for paying for services that had, in many though not all cases, been previously gained extralegally. As Lefebvre (1992) argues, to claim a right to the city is to make a collective argument for a more inclusive, transformed, and renewed city. The concept of being well connected that I have articulated in this book is just such a claim to collective belonging. In Ezba, this claim is based on shared struggle, a broadly articulated but resolutely particular connection between people in a diverse area. The future city that well-connected residents seek recognizes multiple versions of a well-lived life and provides the structures necessary to achieve such visions.

Globally, one of the leading causes of children's death is diarrheal disease from poor water quality, and populations around the world are ever more urban. Therefore, urban water issues, such as quality, price, and access, are key problems for the twenty-first century. In this book I have told the story about how people in one specific place try to make connection happen in potable and wastewater systems, which are imagined to be inherently and universally good. While these water systems can offer improved standards of living, they are not unproblematic changes. They require quite a lot of labor from people to make the connections happen; those connections don't always look like what people anticipated; and such systems can interrupt networks of sociality that those in the urban margins need to survive. I have demonstrated that what's at stake is not just the material conditions and microbiological elements of water, but the social and ethical systems that make places livable and connected. We need to pay attention to everyday water practices in places like Ezbet Khairallah.

In conversations with Ezba residents while watching the progress of installation on the street below, they expressed anxiety about what water system improvements might mean for the neighborhood. Maybe Ezba would be a dream place to live, some hoped, relatively close to downtown and with easy access to transportation routes. In this vision, residents had optimism that their actual lives might live up to the life trajectory that they sought, which they read as healthy and dignified. While residents of Ezba did not expect a life without trials and tribulations, they did have hope that robust social networks and smart social labor could help them weather dips and disruptions.

But if Ezba becomes a dream place to live, will the people who currently live there still be around to enjoy the benefits? How will residents remain well connected to their home and to each other in the face of change? How will they forge and manage new connections with the state in the future, and how far might their life trajectory be sent off track by these larger forces? In short, will Ezba as they know it float away from them? These were the questions that residents of Ezba pondered as we watched the backhoes digging below, water and sewage streaming away down the street.

Notes

CHAPTER ONE: **Introduction**

1. Another joke demonstrates this idea, pointing to the unfair advantages that some enjoy due to their positions of power: *Law kān dimāghhā biyūgaʿhā skīt abūhā yinfaʿhā*. This roughly translates to "If she has a worry/headache, her dad's position will solve it for her."

2. See, for example, the Corruption Perceptions Index (Transparency International 2020) and the 2010 report on Egypt by the UN special rapporteur on the human right to safe drinking water and sanitation (Albuquerque 2010). For more nuanced discussions, see Feldman 2018; Doughan 2018; and Wedeen 2008.

3. See Peterson 2011, in which the author examines the intersection of global media flows and local youths' experiences in Cairo as processes of searching for connection.

4. These informal infrastructures are what James Scott (1990) calls "infrapolitics" and what Partha Chatterjee (2004) calls "political society." Scholars such as Chatterjee, Scott, and others have pushed back against the liberal-illiberal binary of state infrastructures, articulating the mutually constitutive nature of formal and informal mechanisms of governance. They point both to the inaccurate self-description of the liberal idea of the state and to the reality of illiberal politics in Western nations. For example, Anand 2017 demonstrates the mutual constitution of liberal and illiberal practices, showing how the fact that Mumbai's water system is leaky (in terms of both water and authority) actually helps to preserve public water infrastructures. Abdelrahman 2005 shows that the lines between civil society and political society are indistinct at best.

5. Scholarly attention to questions of "good governance" demonstrates the tension between local discontent about predatory, inaccessible, or ineffective governance (Adly 2020; Doughan 2018; Muir and Gupta 2018; Smith 2007) and normative development discourses about the immorality of particular bureaucratic practices (Hetherington 2019; Gupta 2012) and their incompatibility with postcolonial transitions to democracy (Tidey 2022; Pierce 2016).

6. In their meta-analysis of cross-cultural household water-sharing practices, Wutich and colleagues 2018 demonstrate that in informal urban spaces, households share in order to cope with poor supply (irregular supply or not enough altogether) and high costs, in response to religious ideals, and to try to secure and augment personal and political interests. The authors argue that household water sharing is an understudied topic in social science literature and deserves more attention, given that we live in an increasingly urban, hot, and water-scarce world.

7. There is some debate about whether the Nile or the Amazon is the longest river in the world.

8. Mitchell's 2002 work takes on the Nile and many of the iterations of history, technology, force, and fate that help to explain its current form. Egypt has a long history of technologically managing the Nile River: measuring its annual flooding and creating dams and other infrastructure that transport water through a aseries of irrigation canals to capitalize on natural forces to serve human demands. Historically, the Nile floods brought not just an abundance of water, but also a plentitude of rich alluvial mud, some 10 million tons of which were deposited annually on the floodplain, traveling down from the Blue Nile River and contributing to the fecundity of the Egyptian Nile Valley (Mitchell 2002: 195). David Sims, in his 2014 study on desert land reclamation projects in Egypt, argues that development practices in Egypt have not been driven by the stark contrast between the fertile Nile and the barren desert, as the usual story has it. Instead, he says, development in Egypt is due to its unique geography: large tracts of desert border the developed areas along the river valley and both coasts, and the Egyptian state uses this desert area as a vast reserve of empty state-owned land available to endless piecemeal encroachment. See also Derr's 2019 historical monograph on the river as a lived experience, and Barnes 2014 on the sociotechnical production of agricultural water in Egypt.

9. Heinl and Thorweihe 1993 estimated that the Nubian Sandstone Aquifer housed 150,00 square kilometers of water at that time. In 2004, Gossel, Ebraheem, and Wycisk described the difficulties of estimating the extent and condition of the Nubian Sandstone Aquifer.

10. See Verhoeven 2019 for a critique of the supply-demand dichotomy used so often to talk about water.

11. It is important to note that Anand has shifted from the discourse of citizenship at the nation-state level to the discourse of citizenship at the municipal scale, in line with the Marxist cultural geography of Lefebvre 2003 and Harvey 2008, which focuses on the right to the city. Noura Wahby (2018) argues that Egypt is a hydrological state, in that it works to secure its legitimacy and enact governance through water management and provision.

12. A number of scholars have reexamined the received categories of nations, states, civil societies, and communities, looking at what has escaped articulation—the broader spectrum of human effort and lifeworlds that these "dead" categories fail to express. For some researchers, the activities of popular communities provide a strong metric against which to measure the validity of such received categories (Appadurai 2000, 1996; Chatterjee 2004, 1998a, 1998b). Other scholars seek to examine the state from its margins (Rieker and Ali 2008a, 2008b; Gupta and Sharma 2006; Das and Poole 2004), as a phantom (Stamatopoulou-Robbins 2019), or in its everyday manifestations (Ferguson and Gupta 2002; Trouillot 2001; Abrams 1988). Still others use the loose category of informality as a starting place from which to investigate what is silenced in the search for taxonomy. Some of the most compelling of these reenvisionings demand that we pay attention in ways that are awake to the contradictions, rhythms, and flows of people making do (Certeau 2011) and getting along in the creation of novel networks and communities (Simone 2018, 2004a).

13. See Bayat 2013 on the ways in which the pursuit of life chances by urban residents creates political shifts in the Middle East.

14. The sanitary revolution of the nineteenth century was a colonial project designed to manage local populations and maintain the health of colonial authorities (Benidickson 2007). See Anderson 2006 for an example of the relationship between hygiene and colonialism. See Ali 2002 for a look at national and international discourses about bodily regulation as they play out in Egypt.

15. As Farha Ghannam (2002) says of her own fieldwork, my status as a young married woman gave me a great deal of access to homes, but it placed certain limitations on my movements in the public sphere.

CHAPTER TWO: **Sowing Connection**

1. Local naming practices often assign women the honorific Um, meaning "mother of" their eldest child, so Um Wasam is Wasam's mother, and Um Kareem is Kareem's mother. People in a local area will know women by these nicknames and possibly by some other identifying detail. There is flexibility in this practice, as some women prefer to be identified by the name of their eldest child while others prefer to be identified by the name of their eldest male child. The degree of social closeness can be indexed by how people refer to and about others.

2. I had seen the open-backed trucks transporting people into and out of Ezba and had come to the area where I had seen people being loaded. I didn't know how much the ride would cost, or what the unspoken rules were for navigating a seat. Um Wasam and Um Kareem took pity on the newcomer, told me that the cost was half an EGP for adults, and steered me toward the front of the pickup bed closest to the cabin. We chatted during the ride down to Zahraa and made plans to meet the next week.

3. Firm figures on the population and size of the area are difficult to come by because of its informality and the fact that it falls into four different administrative districts within the city (Deboulet 2009).

4. During the 1950s and '60s, the state undertook several subsidized public housing projects, primarily for the lower-income classes. Later, it added upmarket housing, which was often rented to government officials and military officers (Sims 2012: 50–51). Housing privileges—particularly access to newer and more luxurious units—are among the perks that the military has been anxious to maintain in the postrevolutionary period since 2011. The public housing was almost exclusively under rent laws that granted tenure in perpetuity as long was rent was paid (Sims 2012). In 1996, the Egyptian legislature passed a law that exempted new rentals from rent control laws, setting off substantive changes in the rental property markets.

5. This coincided with Sadat's Infitah (open door) policy, which reversed earlier moves by President Gamal Abdel Nasser to redistribute landholdings nationally. In the mid-1970s, Sadat began a program to expand housing into the desert areas surrounding Cairo, and urban poor people began to be channeled into new locations.

6. The Cairo ring road, which was constructed between 1985 and 2001, divides Ezba into two parts. The section of the road that passes by Ezba was constructed in the last phase of the project due in part to conflicts over the possible desecration of Jewish graves in the southern cemeteries. Cutting the ring road through areas such as Ezba and el-Basateen had the unintended consequence of making the vast areas of informal housing

in Cairo visible to the driving public in new ways. The GOC responded by painting the external walls of many of the buildings closest to the road (Tadamun n.d.).

7. To "see without seeing" is a term that one of my interviewees used to describe how the governorate has dealt with areas like Ezbet Khairallah, as I detail in chapter 4.

8. Presidential decrees 1197 (1972) and 1420 (1974).

9. Tourism accounted for roughly 12 percent of Egypt's gross domestic product in fiscal year 2018–2019, and the vast majority of that was from leisure tourism. For a discussion of the ways in which heritage preservation impacts city planning, see Panetta 2020.

10. Globally, this is a common strategy for women who have to carry water to their homes, as violence, including sexual violence, is often a real threat in that process (Hahn, Cless, and Soentgen 2012).

11. Feminist political economist Bina Agarwal (1998) helps to break down the idea of the household as a de facto unit of analysis in economic and development research, showing the ways in which it can hide inequalities in access and labor distribution.

12. Yes, there are monetary costs to these burdens. But articulating them only in those terms overemphasizes productive work in service of creating surplus value, as feminist economists have long pointed out (Waring 1990). The reproductive work that women do to produce family members as socially viable (Elyachar 2005) is, broadly speaking, rendered invisible in the capitalist paradigm. Feminist scholars have sought to make this social labor hypervisible as burdensome and unpleasant barriers to women's full participation in education and labor markets. Even in feminist rereadings that insist on domestic labor as *labor*, the negative and burdensome nature of the work is emphasized because it disables women's self-actualization and rest (labor's opposite). What I realized in talking about the issue of hauling water in Ezbet Khairallah is that women can both be burdened by and also find meaning and joy in domestic labor. They appreciate both the challenges of it and the love and fulfillment that service to others can bring. As scholars Saba Mahmood (2011) and Lila Abu-Lughod (2002) point out, reading this as false consciousness is a poor way to think of the ways in which people construct meaningful lives for themselves.

13. Ali quoted 5 percent, but according to Tadamun n.d., residents of Ezbet Khairallah pay 7 percent of the value of their land to the Governorate of Cairo for the right to use it.

14. One interviewee described having barricaded the road from Zahraa with barrels of flaming oil to prevent police from entering. See Tadamun's (n.d.) history of the area, which is referred to as `Izbit Khayrallah.

15. David Sims (2016) argues that in Egypt, formal land tenure may not be as important as the scholarly literature about urban informality assumes. Sims points out that residents of Cairo are often able to access services and maintain their land despite their lack of legal status. However, Sims does not address the affective dimensions of this informal existence: inhabitants of Ezba feel real anxiety about the uncertainty of their tenure and particularly about the added pressure of moves by the el-Sisi regime to reclaim state land.

16. The demolitions along Establ Antar on the northwestern ridge of Ezba above Zahraa were made under the authority of this classification. See Khalifa 2011 for a discussion of this process at the national level.

17. "Left alone" echoes the notion that when it comes to problematic informal areas like Ezba, the state "sees without seeing," ignoring the settlements' need for services.

State services are often extended to informal areas where residents actually have purchased their land; these areas were classified as informal because the land was originally zoned for agricultural uses rather than for residential or business construction. In these areas, the necessary "paper life" could be established.

18. Pushing decentralized planning is part of the projects of a number of international organizations working in Egypt. The central government's firm hold on the authority to make changes like these is often seen as a block to development.

19. Land-based finance is a policy that is favored by organizations like the World Bank and the IMF, because it provides a non-debt-based way to fund the construction of necessary infrastructure in developing countries. Proponents argue that it encourages efficiency in urban land markets (Peterson 2008); however, one could also argue that it opens up vulnerable areas in the urban Global South to more direct appropriation by moneyed interests.

20. After the 2011 revolution, the plan was put on hold for some time (Ibrahim 2015), but it was picked back up by the el-Sisi government in what Robert Flahive (2018) refers to as "regime security urbanization"—a project to reform urban spaces to reduce the possibility for mass mobilization.

21. Unni Wikan's 1996 work focuses on the lives and living conditions in a popular neighborhood in Cairo and the constant struggle for urban poor people to maintain a sense of self-respect in the face of nearly overwhelming difficulties In a similar vein, Evelyn Early (1993) focuses on health practices in the everyday life of poor neighborhoods in Cairo. Situating herself as a symbolic anthropologist, Early (1993: 24) looks for the "coded references, the nuggets of cultural wisdom uttered in the cracks of ordinary discourse" from which a sense of the cultural worlds of these urban poor women can be deciphered.

22. Asef Bayat (2013) argues that passive networks formed through the recognition of similarities in life situations, self-presentation, and leisure activities play an important role for marginalized urbanites, women, and young people in the Middle East and North Africa as they participate in changing political possibilities. They do so through the accretion of many daily actions, which he says are part of the kinds of urban nonmovements that subvert governing norms and infringe on property and public space.

CHAPTER THREE: **Locating Connection**

1. This is also true in other places, as Stamatopoulou-Robbins 2019 demonstrates. Multiple manifestations of waste and wastewater serve as mechanism of governance in the settler colonial context of Palestine. Political prisoners in Iraq have had their own excrement used as a biopolitical weapon against them (Al-Mohammad 2007).

2. The sister-in-law was likely a relatively new bride, and her apartment would have been decorated and furnished as a key aspect of the marital negotiations. There are ever-increasing expectations on both the bride's and groom's families to demonstrate status through consumer goods.

3. Globally, residents of urban informal areas often agitate unsuccessfully for sewage systems (Jenkins and Scott 2007). As in Ezba, even when sewage systems are implemented, they may fail to function well in a particular community because they do not match its sanitation needs, whether for technical or social reasons (Joshi, Fawcett, and Mannan 2011).

4. The infrastructure for this decision had been laid in the 1990s, when state and international fears about the rise of militant Islamic ideologies, supposedly centered in urban informal areas, prompted an expansion of a variety of services to these areas. Chief among these new services was the construction of police stations, but other projects were also initiated, such as bringing potable water to improve people's living conditions (Sims 2012).

CHAPTER FOUR: **Hedging Connection**

A previous version of portions of this chapter appeared in the *Middle East Law and Governance Journal* (Farmer 2017).

1. In Persian literature, the wise idiot figure often appears under the moniker Mullah Nasruddin.

2. Susan Gal and Judith Irvine (2019) write that language is key to the project of political legitimacy and authority, and we should pay attention to the semiotic projects of political actors rather than accepting statements for their referential or truth value.

3. According to both the USAID and Chemonics officials who work in the water sector in Egypt, "right-sizing" the staff of the water companies is a primary objective, but it is not an easy one. According to the USAID representative, the HCWW needs to reduce staff before it increases tariffs, so that the costs passed on to the consumer are the optimized operating costs. Osgood thought that major improvements had been made in reducing operating costs everywhere but in staffing, and that this was an entrenched problem:

> The holding company has made some good strides on the cost side, in terms of rationalizing costs, achieving better efficiencies. But there is only so far they can go on that. And so they are still plagued with low tariffs, and in Egypt you don't fire people. You can't fire people. So, they occasionally talk about a hiring freeze, but nobody ever sustains a hiring freeze because they can't fire people. When they have a nonperformer, they just shuttle them over to the side, and they hire people to do the work that the nonperformers are supposed to be doing. But the nonperformers are still in the company and still get paid. I mean, that's the management methodology here. So, there are still a number of problems.

4. The fourteen regional water and wastewater companies had been created after the 1981 water sector reforms, which sought to consolidate and improve on a patchwork of projects built with development funds received after Sadat's 1970s institution of the open-door policy. Between 1975 and 1989, USD 1.5 billion in USAID funds were paid to US engineering and construction corporations to build drinking water plants and sewage collection networks in Egypt. These water funds accounted for some 23 percent of the total USAID development (nonmilitary) funds allocated to Egypt during that period (Mitchell 2002: 238–39). The money was spent primarily, though not exclusively, in Cairo and Alexandria, the two oldest existing water companies. These had been established as private entities during the nineteenth century in the centers of colonial government and were nationalized in 1954 by President Nasser. Outside of these cities, water was provided by municipal and governorate systems, with potable water and wastewater systems run by separate entities.

5. In 2016, things had changed. Widespread anger at the large fluctuations in utility costs gave rise to the Twitter hashtag *imsak fatūra* (catch a bill), and many Egyptians posted images of their electricity bills and implicitly called out the state for theft (Howeidy 2016).

CHAPTER FIVE: **Sensing Connection**

1. For most of the twentieth century, a community of people known as the Zabbaleens, primarily Coptic Christians, collected and recycled most of Cairo's garbage. The organic waste was used to feed their pigs. In 2003, municipal authorities awarded contracts to international corporations for trash collection, impeding but not stopping the Zabbaleens' work in collecting trash around the city. In many cases the Zabbaleens were subcontracted by the multinationals to do garbage collection, but there were conflicts over the ownership of the material collected. In 2009, the municipality killed off most of the pigs within the city limits as a precaution during the global H1N1 (swine flu) epidemic. Without pigs to feed, much of the organic waste of the city remained uncollected.

2. Familarity does not necessarily imply better sensory attunement, however, as the description of Ezba given to me by Dr. Magdy Allam, the representative to the People's Assembly for the Zahraa area, demonstrates. In response to my question about the environmental issues in Ezba, he said:

> There are all kinds of issues there. [laughs] You can have the whole list of pollution and put it there. They are lacking sanitary water. They are lacking drainage for the sewage water. They are lacking a solid waste system. They have some air pollution because of the open burning of solid wastes. Because in the slums there are small industries using furnaces, and in the furnaces, they are using some of the plastics, so this is another source. They are mixing activities. The house is not only a house. It is a house and workshop and farm all at the same time, so this is another kind of pollution. The social pollution is a major one. So they are suffering from many environmental problems.

3. The concern about foreign water is connected to the concern about foreign life *within* water. Interspecies relationships are one important aspect of how both the general public and scientists make sense of water quality (Scaramelli 2021, 2013). Christian McMillen (2020a, 2020b) discusses the ways in which changing agricultural practices helped to spread diseases along the Nile, especially through snails (schistosomiasis) and mosquitoes (malaria). Timothy Mitchell (2002) importantly examines the intersection of phosphorus use for weapons during World War II and starvation and the resulting susceptibility to mosquito-borne illness. Work by Jennifer Derr (2019) on the Nile shows that narratives about schistosomiasis as a water-quality problem overlook the ways in which technological and economic projects changed the temporality of water in ways that impacted human health.

4. Spackman and Burlingame 2018 traces the ways in which the professionalization of water management over the twentieth century explicitly privileged analytical water-testing methods and relegated the sensory experiences of potable water consumers to inconsequential aesthetics. Scaramelli (2013: 151), however, argues that technical and sensory measures of water quality are themselves not a binary: "perception and scientific

measurement are not the stable ends of a dichotomy; rather, they interact and produce one another: environmental tests are called to respond to and account for bodily sensations, and sensorial perceptions are re-learned in changing environments."

5. The principle of the general trust for water quality in the United States is observed mainly through the breach, such as the justifiable public anger over lead poisoning in Flint, Michigan (Masten, Davies, and Mcelmurry 2016). The quality of water intended for other beneficial uses (ecosystem maintenance, leisure activities) may come to the public's attention as well.

6. In a project of attunement to the somatic experiences of long-term chemical exposure, Nicholas Shapiro (2015: 374) writes, "Smells, whether off-putting or alluring, are most pronounced at the crossings of thresholds and then, over time, recede from perception as they become incorporated into new sensorial norms. As one's scent sensitivity down-regulates in a process of olfactory adaptation, ongoing and low-level exposures become ordinary and perceptually un-detectable (Dalton and Wysocki 1996)—if such exposures even crested scent detection thresholds in the first instance."

7. For example, technological and sensory experiences of water quality offer overlapping explanatory frameworks for the diverse array of folks with competing claims to rights to the spiritual and economic benefits of the Ganga (Ganges) River, according to Kelly Alley (2002).

8. See Sherine Hamdy's (2008) examination of how poor Egyptians suffering from end-stage renal failure make sense of the multiple vectors of harm that they've experienced, including poor state control of water quality.

9. In her 2010 report on Egypt, UN special rapporteur Albuquerque demonstrated the ways in which statistics such as these present an incomplete picture because they don't take into account the fact that the reliability of service is not measured, water from particular sources is assumed to be of good quality despite evidence of contamination, and the analysis is not disaggregated according to income level (Albuquerque 2010).

10. Many of the wealthy residents of Cairo utilize bottled water to get around the water-quality issues in the city, but as Osgood pointed out, "You never know where the bottled water comes from either, so I don't know [if that helps]." Bottled water is a privileged way to get out of the bad infrastructural system, but it is no guarantee of healthy water. See Wahby 2018.

11. During these demonstrations, residents of Ezbet Khairallah pulled blue water jerricans onto the major thoroughfare of the ring road to draw attention to the water challenges that the area faces. See chapter 2.

12. Research also has shown that although viruses and some parasites do not grow in biofilms, they can be trapped in them (LeChevallier 1999).

13. Shehab Ismail discusses an earlier moment when improvements to potable water infrastructure by the British Colonial Government were rejected by Cairenes due to taste. Like Scarmelli (2013), Ismail troubles neat distinctions between local and scientific epistemologies and focuses on emplaced somatic experience as key to understanding knowledge claims (2018).

14. This is a common solution in Egypt. My own household in Cairo did this, managing to avoid the costly and environmentally detrimental practice of relying on bottled water.

15. See Ballestero 2019b on nonessentialist understandings of sensory perceptions that move beyond embodied and phenomenological approaches.

CHAPTER SIX: **Conclusion**

Portions of this chapter appeared in the *Journal of Sustainability Education* (Farmer et al. 2020).

1. This was an example often repeated in conversations about the value of water. The popular *ḥadīth* tells of a man who went to a well in the desert to get a drink. When he climbed out of the well, he saw a dog that was licking a mirage of water. He climbed back down into the well, carried water up in his shoes by his teeth, and gave the dog the water. God thanked the man and forgave him his sins. This story was often paired with a *ḥadīth* about a woman who was careless and closed a cat up in a building without water or the ability to forage for food on its own, and the death of the cat was the reason for the woman being consigned to hell.

2. Writing about irrigation and drinking water *sabīls* in a southern Nile delta village, Hagar ElDidi and Esteve Corbera (2017) point to the ways in which *sabīls* operate in a moral economy of reciprocity that functions to extend access to water resources. They argue, however, that the prosocial projects of charitable water sharing in this agricultural setting is inevitably asymmetrical due to its embeddedness in a larger political economy of inequality. In a study of household water sharing in Bolivia, Wutich and Ragsdale (2008) document what they call the "moral economy of subsistence," where residents of informal areas work to share resources according to ideas that respond to and exceed market logics; this frame doesn't assume that water-sharing practices are inherently morally positive.

3. This is not an exhaustive list of reasons for creating *sabīls*, and my future research will focus on *sabīls* in Cairo as a form of gifted water.

4. *Ṣadaqa jāriyya* is a voluntary offering of charity or benevolence that is ongoing or continuous. According to a *ḥadīth*, it is one of the three ways in which the balance of good deeds attributed to a person can continue to accumulate after death; the other two are the creation of enduring beneficial knowledge and having virtuous descendants.

5. There was some disagreement about the request to recite the *al-Fātiḥa*, as some viewed any recitation of the Qur'an as undirectable to particular individuals, so of no benefit to the person to whom the *sabīl* was dedicated. Others argued that the request for the *al-Fātiḥa* was really an appeal for kindness toward the deceased loved one; it could mean to read the *al-Fātiḥa* or simply to offer a *duʿāʾ*, mirroring other contexts in which such requests for prayer were made.

6. *Sabīl* users told me that the signs helped them to remember to offer a prayer (often inaudibly) when drinking from a *sabīl*—something that they didn't always remember to do. Several residents later reported that our conversation about *sabīls* had prompted them to offer *duʿāʾ* more often. One reported that she had begun reminding friends when they used *sabīls* together, broadening the network of *khayr* not only by doing good but by reminding others to do good.

7. *Sabīls* as material objects created a number of practical challenges as well. Theft and destruction came up in nearly all of my conversations with *sabīl* keepers—from children playing with and breaking pour spouts, to the theft of the motors from the refrigeration

units of *kūldīrs*, to the use of a broken *ʿulal* in a street fight. One of the key issues was maintenance, since simply keeping *kūldīrs* running with piped water connections, electricity, and mechanical parts was a challenge. *Kūldīrs* were the form of *sabīl* most often abandoned due to unobtainable maintenance.

8. Rural areas also have *sabīls* in the form of standpipes called *trumba* or *ghorab*, one of which is pictured on the cover of this book. See ElDidi and Corbera 2017. These can still occasionally be found in Greater Cairo.

9. The effectiveness of evaporative cooling changes with temperature, humidity, and airflow conditions.

10. The term *zīr* also has been used for the "pot" in pot refrigerator projects (https://en.wikipedia.org/wiki/Pot-in-pot_refrigerator#:~:text=A%20pot%2Din%2Dpot%20refrigerator,which%20the%20food%20is%20placed).

11. When I asked keepers about adding cut fruit, fresh mint, or rose water to their *zīr*, they often referred to a *ḥadīth* about Sayyida ʿAʾisha adding perfume to money that she offered in charity.

12. *Kūlmāns* are much like what you would see on the side of a sports field in the United States. The image on the preceding page is not the same *kūlmān* that I describe here but a very similar one located elsewhere in Ezba.

13. As with the *kūlmān*, the *kūldīr* pictured here is not the *sabīl* being described in the text. In both cases, the quality of the images of these particular *sabīls* were not sufficient for publication, so a similar item from elsewhere in Ezba is depicted.

14. By the time Ezbet Khairallah became a settlement, the tradition of *saqāyīn* (walking water sellers) had largely ended. The *saqāyīn* of Cairo would call out "yiʿawaḍ Allāh" (God will compensate) to draw customers, but the profession faded by the early twenty-first century as piped water systems became more common (Fahmy 2016). Some *sabīls* in Morocco are still referred to as *saqāya* (quenchers of thirst).

References

Abdel Adheem, Ahmed. 2010. "'Slums' Threaten the First Islamic Capital in Africa" [«العشوائيات» تهدد أول عاصمة إسلامية في أفريقيا]. *Al Rai* [جريدة الراي الكويتية], August 28. http://www.alraimedia.com/ar/article/culture/2010/08/28/211565/nr/nc.

Abdel-Baky, Mohamed. 2012. "Victims of the Tap." *Al Ahram Weekly*, no. 1112.

Abd Ellah, Radwan G. 2020. "Water Resources in Egypt and Their Challenges: Lake Nasser Case Study." *Egyptian Journal of Aquatic Research* 46, no. 1 (March 1): 1–12. https://doi.org/10.1016/j.ejar.2020.03.001.

Abdelrahman, Maha M. 2005. *Civil Society Exposed: The Politics of NGOs in Egypt*. London: I. B.Tauris.

Abrams, Philip. 1988. "Notes on the Difficulty of Studying the State (1977)." *Journal of Historical Sociology* 1, no. 1: 58–89. https://doi.org/10.1111/j.1467-6443.1988.tb00004.x.

Abu-Lughod, Janet L. 1971. *Cairo: 1001 Years of the City Victorious*. Princeton, NJ: Princeton University Press.

Abu-Lughod, Lila. 2002. "Do Muslim Women Really Need Saving? Anthropological Reflections on Cultural Relativism and Its Others." *American Anthropologist* 104, no. 3 (September): 783–90. https://doi.org/10.1525/aa.2002.104.3.783.

Adely, Fida. 2009. "Educating Women for Development: The Arab Human Development Report 2005 and the Problem with Women's Choices." *International Journal of Middle East Studies* 41, no. 1: 105–22.

Adly, Amr. 2020. *Cleft Capitalism: The Social Origins of Failed Market Making in Egypt*. Stanford, CA: Stanford University Press.

Afify, Heba. 2014. "Evicted and Nowhere to Go." *Mada Masr*, February 26. https://www.madamasr.com/en/2014/02/26/feature/economy/evicted-and-nowhere-to-go/.

Agarwal, Bina. 1998. "Environmental Management, Equity and Ecofeminism: Debating India's Experience." *Journal of Peasant Studies* 25, no. 4: 55–95.

Albuquerque, Catarina de. 2010. "Report of the Independent Expert on the Issue of Human Rights Obligations Related to Access to Safe Drinking Water and Sanitation." UN Human Rights Council, July 1. http://digitallibrary.un.org/record/685823.

Ali, Kamran Asdar. 2002. *Planning the Family in Egypt: New Bodies, New Selves*. Austin: University of Texas Press.

Allan, Tony. 2001. *The Middle East Water Question*. London: Bloomsbury.

Alley, Kelly D. 2002. *On the Banks of the Ganga: When Wastewater Meets a Sacred River*. Ann Arbor: University of Michigan Press.

Al-Mohammad, Hayder. 2007. "Ordure and Disorder: The Case of Basra and the Anthropology of Excrement." *Anthropology of the Middle East* 2, no. 2 (December 1): 1–23. https://doi.org/10.3167/ame.2007.020202.

Al-Sharqawi, Amira. 2010. "Sit-In of 30 Families in Dar Assalam in Front of the Governorate of Cairo" [اعتصام 30 أسرة بدار السلام أمام محافظة القاهرة." بوابة الأهرام"]. *Al Ahram*, October 18. http://gate.ahram.org.eg/News/7772.aspx.

Amnesty International. 2009. "Buried Alive: Trapped by Poverty and Neglect in Cairo's Informal Settlements." November 12. https://www.refworld.org/docid/4b0262a72.html.

Anand, Nikhil. 2017. *Hydraulic City: Water and the Infrastructures of Citizenship in Mumbai*. Durham, NC: Duke University Press.

———. 2012. "Municipal Disconnect: On Abject Water and Its Urban Infrastructures." *Ethnography* 13, no. 4 (December 1): 487–509. https://doi.org/10.1177/1466138111435743.

———. 2011. "Pressure: The PoliTechnics of Water Supply in Mumbai." *Cultural Anthropology* 26, no. 4: 542–64. https://doi.org/10.1111/j.1548-1360.2011.01111.x.

Anand, Nikhil, Akhil Gupta, and Hannah Appel. 2018. *The Promise of Infrastructure*. Durham, NC: Duke University Press.

Anderson, Warwick. 2006. *Colonial Pathologies: American Tropical Medicine, Race, and Hygiene in the Philippines*. Durham, NC: Duke University Press.

Appadurai, Arjun. 2000. "Grassroots Globalization and the Research Imagination." *Public Culture* 12, no. 1 (January 1): 1–19. https://doi.org/10.1215/08992363-12-1-1.

———. 1996. *Modernity at Large: Cultural Dimensions of Globalization*. 9th ed. Vol. 1. Minneapolis: University of Minnesota Press.

———, ed. 1988. *The Social Life of Things: Commodities in Cultural Perspective*. Cambridge: Cambridge University Press.

Arefin, Mohammed Rafi. 2019. "Infrastructural Discontent in the Sanitary City: Waste, Revolt, and Repression in Cairo." *Antipode* 51, no. 4 (August 1). https://doi.org/10.1111/anti.12562.

Bagir, Zainal Abidin. 2005. *Science and Religion in a Post Colonial World: Interfaith Perspectives*. Adelaide, Australia: ATF Press.

Bakker, Karen J. 2010. *Privatizing Water: Governance Failure and the World's Urban Water Crisis*. Ithaca, NY: Cornell University Press.

Ballestero, Andrea. 2019a. "The Anthropology of Water." *Annual Review of Anthropology* 48 (October 21): 405–21. https://doi.org/10.1146/annurev-anthro-102218-011428.

———. 2019b. *A Future History of Water*. Durham, NC: Duke University Press.

———. 2019c. "Touching with Light, or, How Texture Recasts the Sensing of Underground Water." *Science Technology and Human Values* 44, no. 5 (July 2): 762–85.

Barnes, Jessica. 2014. *Cultivating the Nile: The Everyday Politics of Water in Egypt*. Durham, NC: Duke University Press.

Bayat, Asef. 2013. *Life as Politics: How Ordinary People Change the Middle East*. 2nd ed. Stanford, CA: Stanford University Press.

Bayat, Asef, and Eric Denis. 2000. "Who Is Afraid of Ashwaiyyat? Urban Change and Politics in Egypt." *Environment and Urbanization* 12, no. 2: 185–99.

Beinin, Joel. 2010. *The Struggle for Worker Rights in Egypt*. Washington, DC: Solidarity Center.

———. 2001. *Workers and Peasants in the Modern Middle East.* Cambridge: Cambridge University Press.

Benidickson, Jamie. 2007. *The Culture of Flushing: A Social and Legal History of Sewage.* Vancouver, Canada: University of British Columbia Press.

Björkman, Lisa. 2015. *Pipe Politics, Contested Waters: Embedded Infrastructures of Millennial Mumbai.* Durham, NC: Duke University Press.

Brewis, Alexandra, et al. 2019. "Water Sharing, Reciprocity, and Need: A Comparative Study of Interhousehold Water Transfers in Sub-Saharan Africa." *Economic Anthropology* 6, no. 2: 208–21. https://doi.org/10.1002/sea2.12143.

Brewis, Alexandra, and Amber Wutich. 2019. *Lazy, Crazy, and Disgusting: Stigma and the Undoing of Global Health.* Baltimore, MD: Johns Hopkins University Press.

Bubandt, Nils. 1998. "The Odour of Things: Smell and the Cultural Elaboration of Disgust in Eastern Indonesia." *Ethnos* 61, no. 1: 48–80. https://doi.org/doi:10.1080/00141844.1998.9981564.

Carse, Ashley. 2012. "Nature as Infrastructure: Making and Managing the Panama Canal Watershed." *Social Studies of Science* 42, no. 4 (April 2): 539–63. https://doi.org/10.1177/0306312712440166.

Caton, Steven C. 2017. "Air Pollution as Air Flow: Toward an Ethnography of Smell?" Lecture presented at Harvard University, Cambridge, MA.

Centers for Disease Control and Prevention. n.d. "Global Diarrhea Burden." https://www.cdc.gov/healthywater/global/diarrhea-burden.html.

Certeau, Michel de. 2011. *Practice of Everyday Life.* Berkeley: University of California Press.

Chatterjee, Partha. 2004. *The Politics of the Governed: Reflections on Popular Politics in Most of the World.* New York: Columbia University Press.

———. 1998a. "Beyond the Nation? Or Within?" *Social Text* 16: 57–69.

———. 1998b. "Community in the East." *Economic and Political Weekly* 33, no. 6: 227–82.

Coles, Anne, and Tina Wallace. 2005. *Gender, Water and Development.* London: Routledge.

Cousins, Joshua J. 2017. "Volume Control: Stormwater and the Politics of Urban Metabolism." *Geoforum* 85. https://doi.org/10.1016/j.geoforum.2016.09.020.

Crow, Ben, and Farhana Sultana. 2002. "Gender, Class, and Access to Water: Three Cases in a Poor and Crowded Delta." *Society and Natural Resources* 15, no. 8 (September 1): 709–24. https://doi.org/10.1080/08941920290069308.

Dalton, Pamela, and Charles J. Wysocki. 1996. "The Nature and Duration of Adaptation Following Long-Term Odor Exposure." *Perception and Psychophysics* 58 (July): 781–92. https://doi.org/10.3758/BF03213109.

Darwish, Mahmoud. 2011. *In the Presence of Absence.* Translated by Sinan Antoon. Brooklyn, NY: Archipelago.

Das, Veena, and Deborah Poole. 2004. *Anthropology in the Margins: Comparative Ethnographies.* Albuquerque, NM: SAR Press.

Davis, Mike. 2006. *Planet of Slums.* London: Verso.

Deboulet, Agnès. 2009. "The Dictatorship of the Straight Line and the Myth of Social Disorder: Revisiting Informality in Cairo." In *Cairo Contested: Governance, Urban Space, and Global Modernity*, ed. Diane Singerman. Cairo: American University in Cairo Press.

Deeb, Lara. 2015. "Thinking Piety and the Everyday Together: A Response to Fadil and Fernando." *Hau: Journal of Ethnographic Theory* 5, no. 2: 93–99.

Deeb, Lara, and Jessica Winegar. 2012. "Anthropologies of Arab-Majority Societies." *Annual Review of Anthropology* 41, no. 1: 537–58. https://doi.org/10.1146/annurev-anthro-092611-145947.

Denysenko, Nicholas E. 2012. *The Blessing of Waters and Epiphany: The Eastern Liturgical Tradition*. London: Routledge.

Derr, Jennifer L. 2019. *The Lived Nile: Environment, Disease, and Material Colonial Economy in Egypt*. Stanford, CA: Stanford University Press.

Dhahir, Alaa Al-Din. 2010. "The Crime of 'Al-Saba' Banat" [جريمة «السبع بنات»]. *Masress* [مصرس], July 2. http://www.masress.com/rosadaily/71042.

Doughan, Yazan. 2018. "Corruption, Authority and the Discursive Production of Reform and Revolution in Jordan." PhD diss., University of Chicago. https://doi.org/10.6082/uchicago.1582.

Douglas, Mary. 1966. *Purity and Danger: An Analysis of Concepts of Pollution and Taboo*. New York: Praeger.

Durham, Deborah. 2011. "Disgust and the Anthropological Imagination." *Ethnos* 76, no. 2 (June 1): 131–56. https://doi.org/10.1080/00141844.2010.547941.

Early, Evelyn A. 1993. *Baladi Women of Cairo: Playing with an Egg and a Stone*. Boulder, CO: Lynne Rienner.

ElDidi, Hagar, and Esteve Corbera. 2017. "A Moral Economy of Water: Charity Wells in Egypt's Nile Delta." *Development and Change* 48, no. 1 (January): 121–45. https://doi.org/10.1111/dech.12286.

El Sherif, Heba. 2010. "Polluted Water in Kafr El-Sheikh Threatens Lives, Say Lawyers." *Egypt Independent*, April 4.

Elyachar, Julia. 2010. "Phatic Labor, Infrastructure, and the Question of Empowerment in Cairo." *American Ethnologist* 37, no. 3 (August 1): 452–64. https://doi.org/10.1111/j.1548-1425.2010.01265.x.

———. 2005. *Markets of Dispossession: NGOs, Economic Development, and the State in Cairo*. Durham, NC: Duke University Press.

Fahmy, Ziad. 2016. "An Earwitness to History: Street Hawkers and Their Calls in Early 20th-Century Egypt." *International Journal of Middle East Studies* 48, no. 1 (February): 129–34. https://doi.org/10.1017/S0020743815001531.

Fairhead, James, Melissa Leach, and Ian Scoones. 2012. "Green Grabbing: A New Appropriation of Nature?" *Journal of Peasant Studies* 39, no. 2: 237-61. doi:10.1080/03066150.2012.671770.

FAO. 2016. "AQUASTAT Country Profile: Egypt." Rome, Italy: Food and Agriculture Organization of the United Nations. http://www.fao.org/3/i9729en/I9729EN.pdf.

Farmer, Tessa. 2017. "Willing to Pay: Competing Paradigms about Resistance to Paying for Water Services in Cairo, Egypt." *Middle East Law and Governance Journal* 9: 3–19.

Farmer, Tessa, Mel Throckmorton, and Fazlah Rahaman. 2020. "The Sabils of Cairo: Small Scale Urban Adaptations to Water Stress." *Journal of Sustainability Education* (March 2). http://www.susted.com/wordpress/content/the-sabils-of-cairo-small-scale-urban-adaptations-to-water-stress_2020_03/.

Feldman, Ilana. 2018. "Care and Suspicion: Corruption as Definition in Humanitarian Relations." *Current Anthropology* 59, no. 18: 160–70.

Ferguson, James, and Akhil Gupta. 2002. "Spatializing States: Toward an Ethnography of Neoliberal Governmentality." *American Ethnologist* 29, no. 4 (January 7): 981–1002. https://doi.org/10.1525/ae.2002.29.4.981.

Flahive, Robert. 2018. "Regime-Security Urbanism: Cairo 2050 and Beyond in al-Sisi's Cairo." *Jadaliyya*, July 17. https://www.jadaliyya.com/Details/37753.

Frøystad, Kathinka. 2018. "Failing the Third Toilet Test: Reflections on Fieldwork, Gender and Indian Loos." *Ethnography* (October 15). https://doi.org/10.1177/1466138118804262.

Furniss, Jamie. 2017. "What Type of Problem Is Waste in Egypt?" *Social Anthropology* 25, no. 3 (August): 301–17. https://doi.org/10.1111/1469-8676.12421.

Gal, Susan, and Judith T. Irvine. 2019. *Signs of Difference: Language and Ideology in Social Life*. Cambridge: Cambridge University Press.

Gamal-Eldin, Mohamed. 2020. "Doing Environmental, Infrastructural, and Urban Histories along the Suez Canal." *Jadaliyya*, October 22. https://www.jadaliyya.com/Details/41886.

Gandy, Matthew. 2014. *The Fabric of Space: Water, Modernity, and the Urban Imagination*. Cambridge, MA: MIT Press.

———. 2004. "Rethinking Urban Metabolism: Water, Space and the Modern City." *City* 8, no. 3 (December 1): 363–79. https://doi.org/10.1080/1360481042000313509.

———. 2003. *Concrete and Clay: Reworking Nature in New York City*. Cambridge, MA: MIT Press.

Ghannam, Farha. 2002. *Remaking the Modern: Space, Relocation, and the Politics of Identity in a Global Cairo*. Berkeley: University of California Press.

Gossel, W., A. M. Ebraheem, and P. Wycisk. 2004. "A Very Large Scale GIS-Based Groundwater Flow Model for the Nubian Sandstone Aquifer in Eastern Sahara (Egypt, Northern Sudan and Eastern Libya)." *Hydrogeology Journal* 12, no. 6 (December 1): 698–713. https://doi.org/10.1007/s10040-004-0379-4.

Gouda, Dalia M. 2016. *Social Capital and Local Water Management in Egypt*. Cairo: American University in Cairo Press.

Groenfeldt, David, and Jeremy Schmidt. 2013. "Ethics and Water Governance." *Ecology and Society* 18, no. 1 (February 4). https://doi.org/10.5751/ES-04629-180114.

Gupta, Akhil. 2012. *Red Tape: Bureaucracy, Structural Violence, and Poverty in India*. Durham, NC: Duke University Press.

Gupta, Akhil, and Aradhana Sharma. 2006. "Globalization and Postcolonial States." *Current Anthropology* 47, no. 2 (April): 277–307.

Hafez, Sherine. 2013. "Will the Rational Religious Subject Please Stand Up? Muslim Subjects and the Epistemology of Rationality." In *Anthropology of the Middle East and North Africa: Into the New Millennium*, ed. Sherine Hafez and Susan Slyomovics. Bloomington: Indiana University Press.

———. 2011. *An Islam of Her Own: Reconsidering Religion and Secularism in Women's Islamic Movements*. New York: New York University Press.

Hahn, Hans Peter, Karlheinz Cless, and Jens Soentgen. 2012. *People at the Well: Kinds, Usages and Meaning of Water in a Global Perspective*. Chicago, IL: University of Chicago Press.

Hamdy, Sherine. 2008. "When the State and Your Kidneys Fail: Political Etiologies in an Egyptian Dialysis Ward." *American Ethnologist* 35, no. 4: 553–69. https://doi.org/10.1111/j.1548-1425.2008.00098.x.

Hamdy, Sherine, and Coleman Nye. 2017. *Lissa: A Story about Medical Promise, Friendship, and Revolution*. Toronto: University of Toronto Press.

Han, Clara. 2012. *Life in Debt: Times of Care and Violence in Neoliberal Chile*. Berkeley: University of California Press.

Haraway, Donna J. 2007. *When Species Meet*. Minneapolis: University of Minnesota Press.

Harris, Leila M., Jacqueline A. Goldin, and Christopher Sneddon. 2013. *Contemporary Water Governance in the Global South: Scarcity, Marketization and Participation*. London: Routledge.

Harvey, David. 2010. *Social Justice and the City*. Athens: University of Georgia Press.

———. 2008. "The Right to the City." *New Left Review* 53 (September–October). https://newleftreview.org/issues/ii53/articles/david-harvey-the-right-to-the-city.

———. 2003. "The Right to the City." *International Journal of Urban and Regional Research* 27, no. 4: 939–41. https://doi.org/10.1111/j.0309-1317.2003.00492.x.

Hastrup, Kirsten, and Frida Hastrup. 2015. *Waterworlds: Anthropology in Fluid Environments*. New York: Berghahn.

Hefny, Kamal, M. Samir Farid, and Mohamed Hussein. 1992. "Groundwater Assessment in Egypt." *International Journal of Water Resources Development* 8, no. 2 (June 1): 126–34. https://doi.org/10.1080/07900629208722543.

Heinl, M., and U. Thorweihe. 1993. "Groundwater Resources and Management in SW Egypt." *Environmental Science*. paper/Groundwater-resources-and-management-in-SW-Egypt.-Heinl-Thorweihe/b08e582927782171d7f71409a63b77541c93851f.

Helmreich, Stefan. 2011. "Nature/Culture/Seawater." *American Anthropologist* 113, no. 1 (March): 132–44. https://doi.org/10.1111/j.1548-1433.2010.01311.x.

———. 2009. *Alien Ocean: Anthropological Voyages in Microbial Seas*. Berkeley: University of California Press.

Hendawi, Hamza. 2020. "El Sisi: Nile Dam Must Benefit Egypt, Ethiopia and Sudan Equally." *National*, July 28. https://www.thenational.ae/world/mena/el-sisi-nile-dam-must-benefit-egypt-ethiopia-and-sudan-equally-1.1055942.

Henshaw, Victoria. 2014. *Urban Smellscapes: Understanding and Designing City Smell Environments*. London: Routledge.

Hetherington, Kregg, ed. 2019. *Infrastructure, Environment, and Life in the Anthropocene*. Durham, NC: Duke University Press.

Hirschkind, Charles. 2006. *The Ethical Soundscape: Cassette Sermons and Islamic Counterpublics*. New York: Columbia University Press.

Hoodfar, Homa. 1997. *Between Marriage and the Market: Intimate Politics and Survival in Cairo*. Berkeley: University of California Press.

Howeidy, Amira. 2016. "Protests against Utility Prices Go Online." *Ahram Online*, April 6. http://english.ahram.org.eg/NewsContent/1/0/198920/Egypt/0/Protests.aspx.

Hussein, Abdel-Rahman. 2008. "Emaar Accused of Culpability in Duweiqa Rockslide." *Daily News Egypt*, September 8.

Ibrahim, Kareem. 2015. "Urbanization, Informal Areas and the New Cabinet." *Mada Masr*, September 20. https://www.madamasr.com/en/2015/09/20/opinion/u/urbanization-informal-areas-and-the-new-cabinet/.

Idris, Ali, dir. 2007. *Morgan Ahmed Morgan*. Egypt: Good News.

Ismail, Shehab Fakhry. 2015. "Engineering Heterotopia." *International Journal of Middle East Studies* 47, no. 3 (August): 566–69. https://doi.org/10.1017/S0020743815000574.

Jagannathan, N. Vijay, Ahmed Shawky Mohamed, and Alexander Kremer, eds. 2009. "Water in the Arab World: Management Perspectives and Innovations." Washington, DC: World Bank. http://documents1.worldbank.org/curated/en/590691468052797602/pdf/495930WP0Water1Box341965B01PUBLIC1l.pdf.

Jenkins, Marion W., and Beth Scott. 2007. "Behavioral Indicators of Household Decision-Making and Demand for Sanitation and Potential Gains from Sanitation Marketing in Ghana." *Social Science and Medicine* 62, no. 12 (July): 2427–42. https://doi.org/doi:10.1016/j.socscimed.2007.03.010.

Jewitt, Sarah. 2011. "Geographies of Shit: Spatial and Temporal Variations in Attitudes towards Human Waste." *Progress in Human Geography* 35, no. 5 (October 1): 608–26. https://doi.org/10.1177/0309132510394704.

Joshi, Deepa, Ben Fawcett, and Fouzia Mannan. 2011. "Health, Hygiene and Appropriate Sanitation: Experiences and Perceptions of the Urban Poor." *Environment and Urbanization* (April 13). https://doi.org/10.1177/0956247811398602.

Kaika, Maria. 2004. *City of Flows: Modernity, Nature, and the City*. London: Routledge.

Kaplan, Martha. 2011. "Lonely Drinking Fountains and Comforting Coolers: Paradoxes of Water Value and Ironies of Water Use." *Cultural Anthropology* 26, no. 4: 514–41.

Khalifa, Marwa A. 2011. "Redefining Slums in Egypt: Unplanned versus Unsafe Areas." *Habitat International* 35, no. 1 (January 1): 40–49. https://doi.org/10.1016/j.habitatint.2010.03.004.

Khalil, Deena. 2019. "The Flexible Governance of Water in Cairo's Informal Areas." *Water* 11, no. 8 (August): 1644. https://doi.org/10.3390/w11081644.

———. 2018. "Neoliberal Water Infrastructure, Informality, and the State in Cairo, 1952–2017: A Case Study of Ezbet El-Haggana." PhD diss., University College London.

Kipper, Regina, and Marion Fischer, eds. 2009. "Cairo's Informal Areas: Between Urban Challenges and Hidden Potentials." https://zlinedezine.files.wordpress.com/2011/11/cairosinformalareas_fulltext.pdf.

Larkin, Brian. 2013. "The Politics and Poetics of Infrastructure." *Annual Review of Anthropology* 42, no. 1: 327–43. https://doi.org/10.1146/annurev-anthro-092412-155522.

Latour, Bruno. 2004. *Politics of Nature: How to Bring the Sciences into Democracy*. Cambridge, MA: Harvard University Press.

LeChevallier, Mark W. 1999. "Emerging Pathogens: Names to Know and Bugs to Watch Out For." In National Resource Council, *Identifying Future Drinking Water Contaminants*. Washington, DC: National Academies Press, 136–72.

Leeuwen, Kees van, et al. 2019. "The Challenges of Water Management and Governance in Cities." *Water* 11, no. 6 (June): 1180. https://doi.org/10.3390/w11061180.

Lefebvre, Henri. 2003. *The Urban Revolution*. Minneapolis: University of Minnesota Press.

———. 1992. *The Production of Space*. Translated by Donald Nicholson-Smith. Hoboken, NJ: Wiley-Blackwell.

Leporte, Dominique. 2002. *The History of Shit*. Translated by Rodolphe el-Khoury. Cambridge, MA: MIT Press.

Li, Tania Murray. 2007. *The Will to Improve: Governmentality, Development, and the Practice of Politics*. Durham, NC: Duke University Press.

Limbert, Mandana E. 2010. *In the Time of Oil: Piety, Memory, and Social Life in an Omani Town*. Stanford, CA: Stanford University Press.

Linton, Jamie. 2010. *What Is Water? The History of a Modern Abstraction*. Vancouver: University of British Columbia Press.

Linton, Jamie, and Jessica Budds. 2014. "The Hydrosocial Cycle: Defining and Mobilizing a Relational-Dialectical Approach to Water." *Geoforum* 57 (November 1): 170–80. https://doi.org/10.1016/j.geoforum.2013.10.008.

Lockman, Zachary. 1994. "Imagining the Working Class: Culture, Nationalism and Class Formation in Egypt, 1899–1914." *Poetics Today* 15, no. 2 (Summer): 157–90.

Lupton, Deborah. 2015. "The Pedagogy of Disgust: The Ethical, Moral and Political Implications of Using Disgust in Public Health Campaigns." *Critical Public Health* 25, no. 1 (January 1): 4–14. https://doi.org/10.1080/09581596.2014.885115.

Lyons, Kristina. 2018. "Rivers Have Memory: The (Im)Possibility of Floods and Histories of Urban De- and Reconstruction in the Andean-Amazonian Foothills." *City and Society* 30, no. 3. https://doi.org/10.1111/ciso.12191.

Mahmood, Saba. 2011. *Politics of Piety*. Princeton, NJ: Princeton University Press.

Manalansan, Martin F., IV. 2015. "The Messy Itineraries of Queerness." Society for Cultural Anthropology, July 21. https://culanth.org/fieldsights/the-messy-itineraries-of-queerness.

Masten, Susan J., Simon H. Davies, and Shawn P. Mcelmurry. 2016. "Flint Water Crisis: What Happened and Why?" *Journal of the American Water Works Association* 108, no. 12: 22–34. https://doi.org/10.5942/jawwa.2016.108.0195.

McFarlane, Colin, and Jonathan Silver. 2017. "Navigating the City: Dialectics of Everyday Urbanism." *Transactions: Institute of British Geographers* 42, no. 3 (January 30): 458–71. https://doi.org/10.1111/tran.12175.

McGlotten, Shaka, and Scott Webel. 2016. "Poop Worlds: Material Culture and Copropower (or, Toward a Shitty Turn)." *S&F Online* 13, no. 3. http://sfonline.barnard.edu/traversing-technologies/poop-worlds-material-culture-and-copropower-shaka-mcglotten-scott-webel/.

McKee, Emily. 2016. *Dwelling in Conflict: Negev Landscapes and the Boundaries of Belonging*. Stanford, CA: Stanford University Press.

McMichael, Celia. 2018. "Toilet Talk: Eliminating Open Defecation and Improved Sanitation in Nepal." *Medical Anthropology* 37, no. 4 (May 19): 294–310. https://doi.org/10.1080/01459740.2017.1371150.

McMillen, Christian. 2020a. "'These Findings Confirm Conclusions Many Have Arrived at by Intuition or Common Sense': Water, Quantification and Cost-Effectiveness at the World Bank, ca. 1960 to 1995." *Social History of Medicine* 34, no. 2 (April 19). https://doi.org/10.1093/shm/hkaa006.

———. 2020b. "Water and the Death of Ambition in Global Health, c. 1970–1990." *História, Ciências, Saúde-Manguinhos* 27 (September): 211–30. https://doi.org/10.1590/s0104-59702020000300011.

Meneley, Anne. 2020. "Hope in the Ruins: Seeds, Plants, and Possibilities of Regeneration." *Environment and Planning E: Nature and Space* 4, no. 1: 158–72. https://doi.org/10.1177/2514848620917516.

———. 1999. "Goods and Goodness." *Social Analysis* 43, no. 3: 69–88.

Miller, William Ian. 1998. *The Anatomy of Disgust*. Cambridge, MA: Harvard University Press.

Mitchell, Timothy. 2014. "Introduction: Life of Infrastructure." *Comparative Studies of South Asia, Africa and the Middle East* 34, no. 3 (January 1): 437–39. https://doi.org/10.1215/1089201X-2826013.

———. 2002. *Rule of Experts: Egypt, Techno-Politics, Modernity*. Berkeley: University of California Press.

Mittermaier, Amira. 2019. *Giving to God: Islamic Charity in Revolutionary Times*. Berkeley: University of California Press.

Mohamed, Abd Alwakeel. 2010. "The People of Izbet Khairallah, Istabl Antar, and Batn al-Baqarah Are among the Dead" [أهالي عزبة خير الل واسطبل عنتر وبطن البقرة في عداد الأموات]. *Masress*, March 19. http://www.masress.com/rosadaily/50334.

Mostafa, Saleh Lamei. 1989. "The Cairene Sabil: Form and Meaning." *Muqarnas* 6. https://doi.org/10.2307/1602278.

Moumtaz, Nada. 2021. *God's Property: Islam, Charity, and the Modern State*. Berkeley: University of California Press.

Muehlebach, Andrea. 2019. "On Possession, Use, and Financialized Infrastructure." *Fieldsights*, March 29. https://culanth.org/fieldsights/on-possession-use-and-financialized-infrastructure.

Muir, Sara, and Akhil Gupta. 2018. "Rethinking the Anthropology of Corruption." *Current Anthropology* 59: S4–S15. https://doi.org/10.1086/696161.

Naguib, Nefissa. 2015. "Aesthetics of a Relationship: Women and Water." In *The Social Life of Water*, ed. John R. Wagner. New York: Berghahn.

Neimanis, Astrida. 2017. *Bodies of Water: Posthuman Feminist Phenomenology*. London: Bloomsbury Academic.

Nucho, Joanne Randa. 2016. *Everyday Sectarianism in Urban Lebanon: Infrastructures, Public Services, and Power*. Princeton, NJ: Princeton University Press.

Nussbaum, Martha. 2006. *Hiding from Humanity: Disgust, Shame, and the Law*. Princeton, NJ: Princeton University Press.

Obeydo, Ahmed, and Bessant Maged. 2010. "The Tears of Ein Al-Sirah, the Most Important Lake for Healing, Are Drowning in the Waters of Sewage and Garbage" [دموع عين الصيرة أهم بحيرة للاستشفاء تغرق في مياه الصرف والقمامة!]. *Masress* [مصرس], March 25. http://www.masress.com/elakhbar/1573.

O'Reilly, Kathleen, and Elizabeth Louis. 2014. "The Toilet Tripod: Understanding Successful Sanitation in Rural India." *Health Place* 29 (September): 43–51. https://doi.org/doi:10.1016/j.healthplace.2014.05.007.

Orlove, Ben, and Steven C. Caton. 2010. "Water Sustainability: Anthropological Approaches and Prospects." *Annual Review of Anthropology* 39, no. 1: 401–15. https://doi.org/10.1146/annurev.anthro.012809.105045.

Pahl-Wostl, Claudia, et al. 2007. "Social Learning and Water Resources Management." *Ecology and Society* 12, no. 2. https://www.jstor.org/stable/26267868.

Panetta, Claire. 2020. "'Digging in the Water?': Grassroots Urbanism, the Egyptian State, and the Politics of Heritage-Making in Post-Revolution Cairo." *International Journal of Heritage Studies* 26 (April 1). https://doi.org/10.1080/13527258.2020.1747102.

Perkins, Alisa. 2015. "Muslim Sound, Public Space, and Citizenship Agendas in an American City." *Citizenship Studies* 19, no. 2: 169–83. http://dx.doi.org/10.1080/13621025.2015.1005948.

Peteet, Julie. 2017. *Space and Mobility in Palestine*. Bloomington: Indiana University Press.

Peterson, George E. 2008. "Unlocking Land Values to Finance Urban Infrastructure: Land-Based Financing Options for Cities." *Gridlines*, no. 40: 4.

Peterson, Mark Allen. 2011. *Connected in Cairo: Growing Up Cosmopolitan in the Modern Middle East*. Bloomington: Indiana University Press.

Pierce, Steven. 2016. *Moral Economies of Corruption: State Formation and Political Culture in Nigeria*. Durham, NC: Duke University Press.

Pierce, Suzanne A., et al. 2013. "Aquifer-Yield Continuum as a Guide and Typology for Science-Based Groundwater Management." *Hydrogeology Journal* 21, no. 2 (March 1): 331–40. https://doi.org/10.1007/s10040-012-0910-y.

Piper, Karen. 2014. *The Price of Thirst: Global Water Inequality and the Coming Chaos*. Minneapolis: University of Minnesota Press.

Povinelli, Elizabeth A. 2002. *The Cunning of Recognition: Indigenous Alterities and the Making of Australian Multiculturalism*. Durham, NC: Duke University Press.

Rademacher, Anne. 2011. *Reigning the River: Urban Ecologies and Political Transformation in Kathmandu*. Durham, NC: Duke University Press.

Ranganathan, Malini. 2015. "Storm Drains as Assemblages: The Political Ecology of Flood Risk in Post-Colonial Bangalore." *Antipode* 47, no. 5 (November 1): 1300–1320. https://doi.org/10.1111/anti.12149.

Redfield, Peter, and Steven Robins. 2016. "An Index of Waste: Humanitarian Design, 'Dignified Living' and the Politics of Infrastructure in Cape Town." *Anthropology Southern Africa* 39, no. 2 (June 10): 145–62. https://doi.org/10.1080/23323256.2016.1172942.

Ribot, Jesse C., and Nancy Lee Peluso. 2009. "A Theory of Access." *Rural Sociology* 68, no. 2: 153–81. https://doi.org/10.1111/j.1549-0831.2003.tb00133.x.

Rieker, Martina, and Kamran Asdar Ali. 2008a. *Gendering Urban Space in the Middle East, South Asia, and Africa*. New York: Palgrave Macmillan.

———. 2008b. *Urban Margins: Envisioning the Contemporary Global South*. Durham, NC: Duke University Press.

Rignall, Karen. 2019. "Living Climate Change in the Middle East and North Africa." *International Journal of Middle East Studies* 51, no. 4: 629–32. doi:10.1017/S0020743819000710.

Rubaii, Kali. Forthcoming. *Counterinsurgency: The Ecology of Coercion*. Durham, NC: Duke University Press.

Samir, Samar. 2020. "Egypt to Impose Harsher Penalty for Water Wastage." *Egypt Today*, July 23. http://www.egypttoday.com/Article/1/90030/Egypt-to-impose-harsher-penalty-for-water-wastage.

Sayed, Ebtehal, et al. 2020. "Sustainable Groundwater Management in Arid Regions: Considering Climate Change Impacts in Moghra Region, Egypt." *Groundwater for Sustainable Development* 11 (October 1). https://doi.org/10.1016/j.gsd.2020.100385.

Scaramelli, Caterina. 2021. *How to Make a Wetland: Water and Moral Ecology in Turkey*. Stanford, CA: Stanford University Press.

———. 2013. "Making Sense of Water Quality: Multispecies Encounters on the Mystic River." *Worldviews: Global Religions, Culture, and Ecology* 17, no. 2 (January 1): 150–60. https://doi.org/10.1163/15685357-01702006.

Schafran, Alex, Matthew Noah Smith, and Stephen Hall. 2020a. *Seeing like a System: The Spatial Contract*. Manchester, England: Manchester University Press.

———. 2020b. *The Spatial Contract: A New Politics of Provision for an Urbanized Planet*. Manchester, England: Manchester University Press.

Schmidt, Jeremy J. 2017. *Water: Abundance, Scarcity, and Security in the Age of Humanity*. New York: New York University Press.

Scott, James C. 1998. *Seeing like a State: How Certain Schemes to Improve the Human Condition Have Failed*. New Haven, CT: Yale University Press.

———. 1990. *Domination and the Arts of Resistance: Hidden Transcripts*. New Haven, CT: Yale University Press.

Sellman, Johanna. 2018. "A Global Postcolonial: Contemporary Arabic Literature of Migration to Europe." *Journal of Postcolonial Writing* 54, no. 6 (November 2): 751–65. https://doi.org/10.1080/17449855.2018.1555207.

Seremetakis, C. Nadia. 1994. *The Senses Still*. Chicago, IL: University of Chicago Press.

Shapiro, Nicholas. 2015. "Attuning to the Chemosphere: Domestic Formaldehyde, Bodily Reasoning, and the Chemical Sublime." *Cultural Anthropology* 30, no. 3 (August 10): 368–93. https://doi.org/10.14506/ca30.3.02.

Simone, AbdouMaliq. 2018. *Improvised Lives: Rhythms of Endurance in an Urban South*. Cambridge: Polity.

———. 2004a. *For the City Yet to Come: Changing African Life in Four Cities*. Durham, NC: Duke University Press.

———. 2004b. "People as Infrastructure: Intersecting Fragments in Johannesburg." *Public Culture* 16, no. 3: 407–29.

Sims, David. 2016. "Securing Land Tenure in Egypt: Who Needs Registered Titles?" *Metropolitics*, June 14. https://metropolitics.org/Securing-Land-Tenure-in-Egypt-Who-Needs-Registered-Titles.html.

———. 2014. *Egypt's Desert Dreams: Development or Disaster?* Oxford: Oxford University Press.

———. 2012. *Understanding Cairo: The Logic of a City Out of Control*. Cairo: American University in Cairo Press.

Singerman, Diane. 2009. "Introduction." In *Cairo Contested: Governance, Urban Space, and Global Modernity*, ed. Diane Singerman. Cairo: American University in Cairo Press.

———. 1996. *Avenues of Participation: Family, Politics, and Networks in Urban Quarters of Cairo*. Princeton, NJ: Princeton University Press.

Singerman, Diane, and Paul Amar. 2006. “Introduction: Contesting Myths, Critiquing Cosmopolitianism, and Creating the New Cairo School of Urban Studies.” In *Cairo Cosmopolitan: Politics, Culture, and Urban Space in the New Middle East*, ed. Diane Singerman and Paul Amar. Cairo: American University in Cairo Press.

Singerman, Diane, and Homa Hoodfar. 1996. *Development, Change and Gender in Cairo: A View from the Household*. Bloomington: Indiana University Press.

Smith, Daniel Jordan. 2007. *A Culture of Corruption: Everyday Deception and Popular Discontent in Nigeria*. Princeton, NJ: Princeton University Press.

Smith, David M. 2000. *Moral Geographies: Ethics in a World of Difference*. Edinburgh: Edinburgh University Press.

Spackman, Christy, and Gary Burlingame. 2018. “Sensory Politics: The Tug-of-War between Potability and Palatability in Municipal Water Production.” *Social Studies of Science* 48, no. 3: 350–71. https://doi.org/10.1177/0306312718778358.

Stamatopoulou-Robbins, Sophia. 2019. *Waste Siege: The Life of Infrastructure in Palestine*. Stanford, CA: Stanford University Press.

Star, Susan Leigh. 1999. “The Ethnography of Infrastructure.” *American Behavioral Scientist* 43, no. 3 (November 1): 377–91. https://doi.org/10.1177/00027649921955326.

Stoller, Paul. 1989. *The Taste of Ethnographic Things*. Philadelphia: University of Pennsylvania Press.

Strang, Veronica. 2004. *The Meaning of Water*. London: Routledge.

Sultana, Farhana. 2011. “Suffering for Water, Suffering from Water: Emotional Geographies of Resource Access, Control and Conflict.” In “New Feminist Political Ecologies,” special issue, *Geoforum* 42, no. 2 (March 1): 163–72. https://doi.org/10.1016/j.geoforum.2010.12.002.

———. 2009. “Fluid Lives: Subjectivities, Gender and Water in Rural Bangladesh.” *Gender, Place, and Culture* 16, no. 4 (July 14): 427–44.

Sultana, Farhana, and Alex Loftus. 2019. *Water Politics: Governance, Justice and the Right to Water*. London: Routledge.

Swyngedouw, Erik. 2015. *Liquid Power: Contested Hydro-Modernities in Twentieth-Century Spain*. Cambridge, MA: MIT Press.

Tadamun. 2015. “Cairo 2050 Revisited: Cairo’s Mysterious Planning: What We Know and What We Do Not Know about Al-Fustat.” September 3. https://medium.com/@tadamun/cairo-2050-revisited-cairo-s-mysterious-planning-what-we-know-and-what-we-do-not-know-about-al-7bc70ebfa3de.

———. n.d. “Tadamun: The Cairo Urban Solidarity Initiative.” http://www.tadamun.co/?post_type=city&p=2741&lang=en&lang=en#.XaSCriV7n6Y.

Tam, Stephanie. 2013. “Sewerage’s Reproduction of Caste: The Politics of Coprology in Ahmedabad, India.” *Radical History Review* 2013, no. 116: 5–30. https://doi.org/10.1215/01636545-1965675.

Thrift, Nigel. 2004. “Intensities of Feeling: Towards a Spatial Politics of Affect.” *Geografiska Annaler: Human Geography*, ser. B, 86, no. 1: 57–78. https://doi.org/10.1111/j.0435-3684.2004.00154.x.

Tidey, Sylvia. 2022. *Ethics for the Right Thing? Corruption and Care in the Age of Good Governance*. Chicago, IL: University of Chicago Press.

Transparency International. 2020. “Corruption Perceptions Index.” https://www.transparency.org/en/cpi/2020.

Tronto, Joan C. 1998. "An Ethic of Care." *Generations: Journal of the American Society on Aging* 22, no. 3: 15–20.

Trouillot, Michel-Rolph. 2001. "The Anthropology of the State in the Age of Globalization." *Current Anthropology* 42, no. 1 (February): 125–38. https://doi.org/10.1086/318437.

Tsing, Anna Lowenhaupt. 2005. *Friction: An Ethnography of Global Connection*. Princeton, NJ: Princeton University Press.

Verhoeven, Harry. 2019. "The Mirage of Supply-Side Development: The Hydraulic Mission and the Politics of Agriculture and Water in the Nile Basin." *Oxford Research Encyclopedia of Environmental Science* (December 23). https://doi.org/10.1093/acrefore/9780199389414.013.645.

Von Schnitzler, Antina. 2013. "Traveling Technologies: Infrastructure, Ethical Regimes, and the Materiality of Politics in South Africa." *Cultural Anthropology* 28, no. 4 (November): 670–93. https://doi.org/10.1111/cuan.12032.

———. 2008. "Citizenship Prepaid: Water, Calculability, and Techno-Politics in South Africa." *Journal of Southern African Studies* 34, no. 4 (December 1): 899–917. https://doi.org/10.1080/03057070802456821.

Wagner, John R. 2015. *The Social Life of Water*. New York: Berghahn.

Wahby, Noura. 2021. "Urban Informality and the State: Repairing Cairo's Waters through *Gehood Zateya*." *Environment and Planning E: Nature and Space* 4, no. 3 (June): 696–717. https://doi.org/10.1177/25148486211025262.

———. 2018. "The Role of the State in Urban Development: The Case of Urban Waterscapes in Cairo, Egypt." PhD diss., Center of Development Studies, University of Cambridge.

Waring, Marilyn. 1990. *If Women Counted: A New Feminist Economics*. Rev. ed. New York: HarperCollins.

Wedeen, Lisa. 2015. *Ambiguities of Domination: Politics, Rhetoric, and Symbols in Contemporary Syria*. Chicago, IL: University of Chicago Press.

———. 2008. *Peripheral Visions: Publics, Power, and Performance in Yemen*. Chicago, IL: University of Chicago Press.

Wikan, Unni. 1996. *Tomorrow, God Willing: Self-Made Destinies in Cairo*. Chicago, IL: University of Chicago Press.

Winegar, Jessica. 2020. "Love and Disgust: Sovereignty Struggles in Egypt's Uprising and After." Lecture at the University of Virginia, January 31.

———. 2012. "The Privilege of Revolution: Gender, Class, Space, and Affect in Egypt." *American Ethnographist* 39, no. 1 (February): 67–70. https://doi.org/doi:10.1111/j.1548-1425.2011.01349.x.

Wutich, Amber, et al. 2018. "Household Water Sharing: A Review of Water Gifts, Exchanges, and Transfers across Cultures." *Wiley Interdisciplinary Reviews: Water* 5, no. 6 (September 7). https://doi.org/10.1002/wat2.1309.

Wutich, Amber, and Kathleen Ragsdale. 2008. "Water Insecurity and Emotional Distress: Coping with Supply, Access, and Seasonal Variability of Water in a Bolivian Squatter Settlement." *Social Science and Medicine* 67, no. 12 (December): 2116–25. https://doi.org/doi:10.1016/j.socscimed.2008.09.042.

Zeiderman, Austin. 2016. "Submergence: Precarious Politics in Colombia's Future Port-City." *Antipode* 48, no. 3: 809–31. https://doi.org/10.1111/anti.12207.

Index

www.ingramcontent.com/pod-product-compliance
Ingram Content Group UK Ltd.
Pitfield, Milton Keynes, MK11 3LW, UK
UKHW041943200825
462084UK00003B/192

9 781421 445489